Algebra 1

LARSON
BOSWELL
KANOLD
STIFF

Applications • Equations • Graphs

Chapter 7 Resource Book

The Resource Book contains the wide variety of blackline masters available for Chapter 7. The blacklines are organized by lesson. Included are support materials for the teacher as well as practice, activities, applications, and assessment resources.

Contributing Authors

The authors wish to thank the following individuals for their contributions to the Chapter 7 Resource Book.

Rita Browning
Linda E. Byrom
José Castro
Christine A. Hoover
Carolyn Huzinec
Karen Ostaffe
Jessica Pflueger
Barbara L. Power
James G. Rutkowski
Michelle Strager

ISBN: 0-618-02045-4

15 14 13 12 11 10 -CKI- 06 05 04

Contents

7 *Systems of Linear Equations and Inequalities*

Contents

Contents

Descriptions of Resources

This Chapter Resource Book is organized by lessons within the chapter in order to make your planning easier. The following materials are provided:

Tips for New Teachers These teaching notes provide both new and experienced teachers with useful teaching tips for each lesson, including tips about common errors and inclusion.

Parent Guide for Student Success This guide helps parents contribute to student success by providing an overview of the chapter along with questions and activities for parents and students to work on together.

Prerequisite Skills Review Worked-out examples are provided to review the prerequisite skills highlighted on the Study Guide page at the beginning of the chapter. Additional practice is included with each worked-out example.

Strategies for Reading Mathematics The first page teaches reading strategies to be applied to the current chapter and to later chapters. The second page is a visual glossary of key vocabulary.

Lesson Plans and Lesson Plans for Block Scheduling This planning template helps teachers select the materials they will use to teach each lesson from among the variety of materials available for the lesson. The block-scheduling version provides additional information about pacing.

Warm-Up Exercises and Daily Homework Quiz The warm-ups cover prerequisite skills that help prepare students for a given lesson. The quiz assesses students on the content of the previous lesson. (Transparencies also available)

Activity Support Masters These blackline masters make it easier for students to record their work on selected activities in the Student Edition.

Alternative Lesson Openers An engaging alternative for starting each lesson is provided from among these four types: ***Application, Activity, Graphing Calculator,*** or ***Visual Approach.*** (Color transparencies also available)

Graphing Calculator Activities with Keystrokes Keystrokes for four models of calculators are provided for each Technology Activity in the Student Edition, along with alternative Graphing Calculator Activities to begin selected lessons.

Practice A, B, and C These exercises offer additional practice for the material in each lesson, including application problems. There are three levels of practice for each lesson: A (basic), B (average), and C (advanced).

Contents

Reteaching with Practice These two pages provide additional instruction, worked-out examples, and practice exercises covering the key concepts and vocabulary in each lesson.

Quick Catch-Up for Absent Students This handy form makes it easy for teachers to let students who have been absent know what to do for homework and which activities or examples were covered in class.

Cooperative Learning Activities These enrichment activities apply the math taught in the lesson in an interesting way that lends itself to group work.

Interdisciplinary Applications/Real-Life Applications Students apply the mathematics covered in each lesson to solve an interesting interdisciplinary or real-life problem.

Math and History Applications This worksheet expands upon the Math and History feature in the Student Edition.

Challenge: Skills and Applications Teachers can use these exercises to enrich or extend each lesson.

Quizzes The quizzes can be used to assess student progress on two or three lessons.

Chapter Review Games and Activities This worksheet offers fun practice at the end of the chapter and provides an alternative way to review the chapter content in preparation for the Chapter Test.

Chapter Tests A, B, and C These are tests that cover the most important skills taught in the chapter. There are three levels of test: A (basic), B (average), and C (advanced).

SAT/ACT Chapter Test This test also covers the most important skills taught in the chapter, but questions are in multiple-choice and quantitative-comparison format. (See *Alternative Assessment* for multi-step problems.)

Alternative Assessment with Rubrics and Math Journal A journal exercise has students write about the mathematics in the chapter. A multi-step problem has students apply a variety of skills from the chapter and explain their reasoning. Solutions and a 4-point rubric are included.

Project with Rubric The project allows students to delve more deeply into a problem that applies the mathematics of the chapter. Teacher's notes and a 4-point rubric are included.

Cumulative Review These practice pages help students maintain skills from the current chapter and preceding chapters.

CHAPTER 7

Tips for New Teachers

For use with Chapter 7

LESSON 7.1

TEACHING TIPS In order to solve a linear system by graphing, students must be very comfortable graphing lines. You might want to spend some time reviewing how to graph lines. Students should be able to graph lines by either using the slope-intercept form of the line or by finding its intercepts. The first system of equations you solve with your students could include one line in standard form (easy to graph by finding its intercepts) and the other line in slope-intercept form. Let students discuss which is the most convenient method to graph each equation.

TEACHING TIPS Once your students are comfortable solving systems by graphing, give them some examples so that they can discover some of the cons of this method. For instance, give students a system such as

$$\begin{cases} y = 2x + 3 \\ y = 3x + 6 \end{cases}, \text{ solution: } (-3, -3)$$

where the graphs of the lines intersect at a very small angle because the lines are almost parallel. Finding the exact point of intersection of these lines can be tricky, because students must be very careful drawing their lines. You can also show students that the graphing method does not work well when the solution is not a lattice point. For example, the following system

$$\begin{cases} y = 7x - 1 \\ y = -2x + 4 \end{cases}, \text{ solution: } \left(\tfrac{5}{9}, 2\tfrac{8}{9}\right)$$

is almost impossible to solve by graphing because the solutions are rational numbers.

COMMON ERROR Some students might check their answer in only one of the equations or not at all. If you used either of the two examples in the previous paragraph, you can show your students how inaccurate the graphing method can be. Remind your students that to make sure they get the right answer, they must check it in *both* equations.

LESSON 7.2

TEACHING TIPS If you did an example in Lesson 7.1 of a linear system where the solutions are not integers, you can use that same example to teach your students how to use substitution to solve a system. This will allow students to compare the two methods.

COMMON ERROR You might want to review how to solve for a variable multiplied by -1, such as x in the equation $-x + 4y = 3$. Otherwise, some students might forget everything about the negative sign and incorrectly solve it as $x = -4y + 3$.

TEACHING TIPS After solving several systems by substitution, give your students some linear systems and ask them *not* to solve them, but to develop an "action plan" for solving them. Their "action plan" should state what variable they would solve for first and in what equation. You can ask them to support their answers with reasons. Show your students that taking a minute to think before starting to solve the system can actually save them time.

TEACHING TIPS You can finish this lesson with a system such as

$$\begin{cases} 3x - 4y = 5 \\ -5x + 4y = 3 \end{cases}, \text{ solution: } (1, 2)$$

While solving this system by substitution requires using fractions, it is a very simple system to solve by elimination. You can use this example to link this lesson with the following one.

LESSON 7.3

COMMON ERROR Some students might try to eliminate one of the variables without checking first whether both equations are already in standard form. Include some examples in your class work where the students must first rearrange at least one of the equations before they can start solving the system.

TEACHING TIPS As in the previous lesson, you can ask students to develop an "action plan" before they actually solve each system. They should state what variable they would eliminate and why. Then they could list the steps needed so that the variable they choose cancels out when they add the equations. Again, emphasize the need to spend some time thinking about the problem before jumping into it.

CHAPTER 7 CONTINUED

Tips for New Teachers

For use with Chapter 7

LESSON 7.3 (CONT.)

COMMON ERROR Some students will multiply the coefficient of the variable they want to eliminate and forget to multiply the other variable and constant of the equation. Remind students that they need to multiply each term of the equation, as if they were using the distributive property with the whole equation.

LESSON 7.4

TEACHING TIPS Start the lesson by asking your students to list pros and cons for each of the three methods they know to solve a linear system of equations. Then ask them to explain how they would decide what method to use for a given system. You can give them several systems and ask them to choose an appropriate method to solve each of them. They should also explain their answers.

LESSON 7.5

TEACHING TIPS Students can learn the number of solutions of special systems by discovery. Ask them to complete Activity 7.5 on page 425, or at least the part titled "Exploring the Concept." Then, have a class discussion where students share and explain their findings.

TEACHING TIPS You might want to introduce and explain the meaning of the terms *consistent*, *inconsistent*, *dependent*, and *independent* in this lesson. If you do so, make sure that students understand their meaning and how they can be combined to describe the nature of a linear system. Check for understanding by asking them questions such as "Can a system be inconsistent and dependent? Why?"

TEACHING TIPS Students do not need to graph a system of linear equations to find out how many solutions it has. All they need to do is to write both equations in slope-intercept form. If the lines have the same slope but different y-intercepts there is no solution, because they are parallel. If the lines have the same slope and the same y-intercept there are infinitely many solutions, because they coincide. Finally, if the lines have different slopes, they intersect each other once and, therefore, there is exactly one solution.

LESSON 7.6

INCLUSION If you have not introduced the concept of *intersection* yet, you need to do so now. Show the connection between intersection and *and* statements such as the ones already covered in compound inequalities. You can refer back to Lesson 6.3, Compound Inequalities, for resources and ideas.

COMMON ERROR Some students might be able to graph the solution region for a given system of linear inequalities, without knowing the significance of doing that. To make sure that students understand why we find the solution region, ask them to make a list of ordered pairs that are possible solutions of the system. They should make the connection that these ordered pairs must be inside the solution region.

Outside Resources

BOOKS/PERIODICALS

Christina, Mary Ann. "Building a Teenage Dance Club." *Mathematics Teaching in the Middle School* (September 1998); pp. 26–30.

SOFTWARE

Rosenberg, Jon. *Math connections: Algebra I.* Pleasantville, NY; Sunburst.

ACTIVITIES/MANIPULATIVES

Jabon, David, Gail Nord, Bryce W. Wilson, Penny Coffman, and John Nord. "Medical Applications of Systems of Linear Equations. *Mathematics Teacher* (May 1996); pp. 398–402, 408–410.

Van Dyke, Frances. "Visualizing Cost, Revenue, and Profit." *Activities: Mathematics Teacher* (September 1998); pp. 488–493, 500–503.

NAME ______________________________ DATE __________

Parent Guide for Student Success

For use with Chapter 7

Chapter Overview One way that you can help your student succeed in Chapter 7 is by discussing the lesson goals in the chart below. When a lesson is completed, ask your student to interpret the lesson goals for you and to explain how the mathematics of the lesson relates to one of the key applications listed in the chart.

Lesson Title	*Lesson Goals*	*Key Applications*
7.1: Solving Linear Systems by Graphing	Solve a system of linear equations by graphing. Model a real-life problem using a linear system.	• Internet • Comparing Cars • Coastal Population
7.2: Solving Linear Systems by Substitution	Use substitution to solve a linear system. Model a real-life situation using a linear system.	• Museum Admissions • Softball Sizes • Running
7.3: Solving Linear Systems by Linear Combinations	Use linear combinations to solve a system of linear equations. Model a real-life problem using a system of linear equations.	• Weight of Gold • Honey Bee Paths • Steamboat Speed
7.4: Applications of Linear Systems	Choose the best method to solve a system of linear equations. Use a system to model real-life problems.	• Career Decisions • Party Planning • Housing
7.5: Special Types of Linear Systems	Identify linear systems as having one solution, no solution, or infinitely many solutions. Model real-life problems using a linear system.	• Statue Heights • Jewelry • Carpentry Supplies
7.6: Solving Systems of Linear Inequalities	Solve a system of linear inequalities by graphing. Use a system of linear inequalities to model a real-life situation.	• Theatre Lighting • Food Budget • Local Magazine

Study Strategy

Pace Yourself. Spend no more than a few minutes on each question. If a question is too difficult, skip it and go back to it if you have time. You can help your student work on pacing by timing some practice tests and then discussing how they went. To aid discussion, you may wish to take notes on what your student has completed after certain intervals.

NAME ______________________________ DATE ____________

Parent Guide for Student Success

For use with Chapter 7

Key Ideas **Your student can demonstrate understanding of key concepts by working through the following exercises with you.**

Lesson	Exercise
7.1	Decide whether the ordered pair $(-2, -3)$ is a solution of the system of linear equations. $7x - 2y = -8$ $3x - 5y = -9$
7.2	Each day you exercise you either walk 5 miles or jog 2 miles. You exercised 19 days last month and the total distance you traveled was 74 miles. How many days did you walk and how many days did you jog?
7.3	Use linear combinations to solve the system of linear equations. $3x - 7y = 16$ $2x + 5y = 1$
7.4	In January, a gas bill is \$235 and the electric bill is \$85. The gas bill decreases \$32 a month from January to August and the electric bill increases \$15 a month during the same period. During which month should both bills be about the same?
7.5	Tell how many solutions the system has. $2x - 10y = 8$ $3x - 15y = 12$
7.6	Give the coordinates of each vertex of the solution region of the system of linear inequalities. $x + 2y \leq 14$ $x \geq 0$ $y \leq 2 + 2x$ $y \geq 0$

Home Involvement Activity

You will need: A paper cup, two measuring cups, and a pin

Directions: Place one measuring cup under a slowly dripping faucet. Record the amount of the water in the measuring cup after one minute and after two minutes. Use the pin to make a small hole near the bottom of the paper cup. Measure one cup of water and pour into the paper cup. Set the paper cup so it drains into the other empty measuring cup. Record the amount of water in the second measuring cup after one minute and after two minutes. Subtract to find the amount of water remaining in the first paper cup after each time interval. If the dripping and the draining started at the same time, when would the first measuring cup and the paper cup have the same amount of water? Have each person guess. Write and solve a system of linear equations to determine when the amounts of water should be the same. Who made the best guess?

Answers

7.1: no ***7.2:*** walked 12 days and jogged 7 days ***7.3:*** $(3, -1)$ ***7.4:*** April ***7.5:*** infinitely many solutions ***7.6:*** $(2, 6)$, $(0, 2)$, $(0, 0)$, $(14, 0)$

NAME ______________________________ DATE __________

Prerequisite Skills Review

For use before Chapter 7

EXAMPLE 1 Simplifying by Combining Like Terms

Simplify the expression.

$9 - 3(5 - x)$

SOLUTION

$9 - 3(5 - x) = 9 + (-3)(5 - x)$	Rewrite as an addition expression.
$= 9 + [(-3)(5) + (-3)(-x)]$	Distribute the -3.
$= 9 + (-15) + (3x)$	Multiply.
$= -6 + 3x$	Combine like terms and simplify.

Exercises for Example 1

Simplify the expression.

1. $8r + (-64r)$
2. $\frac{1}{4}g - \frac{3}{16}g$
3. $\frac{9}{10}s + 2\frac{1}{5}s$
4. $0.1t - 0.02t$

EXAMPLE 2 Solving Equations with Variables on Both Sides

Solve the equation if possible.

$9x + 34 = -x - 6$

SOLUTION

$9x + 34 = -x - 6$	Write original equation.
$9x + 34 + x = -x - 6 + x$	Add x to each side.
$10x + 34 = -6$	Simplify.
$10x = -40$	Subtract 34 from both sides.
$x = -4$	Simplify.

The solution is -4.

When you check the solution, substitute -4 for each x in the equation.

Exercises for Example 2

Solve the equation if possible.

5. $3x + 7(x - 1) = 23$
6. $8x + 4x = 12x - 1$
7. $1 + 7x - 3.1 = 0$
8. $12(8 - y) = 11y + y + 16$

NAME ______________________ DATE __________

Prerequisite Skills Review

For use before Chapter 7

EXAMPLE 3 Verifying Solutions of an Equation

Decide whether the given ordered pair is a solution of the equation or inequality.

a. $5x - y = 15$ $(1, -1)$ **b.** $8y - 4x < 68$ $(3.1, 10)$

SOLUTION

a. The point $(1, -1)$ is not on the graph of $5x - y = 15$. This means that $(1, -1)$ is not a solution.

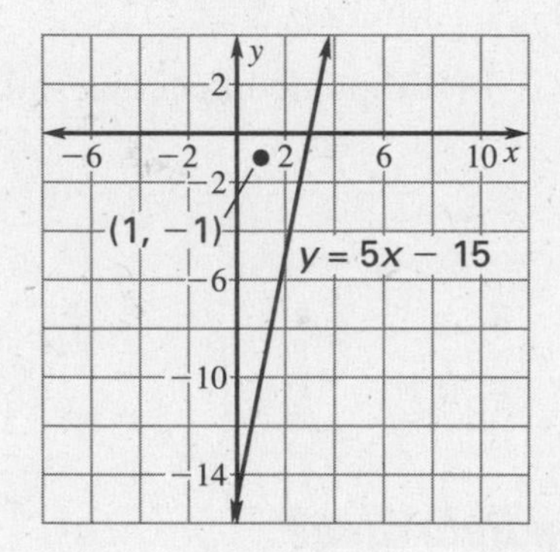

You can also check this algebraically.

$5x - y = 15$	Write original equation.
$5(1) - (-1) = 15$	Substitute 1 for x and -1 for y.
$5 + 1 = 15$	Simplify.
$6 \neq 15$	Not a true statement.

$(1, -1)$ is not a solution of the equation $5x - y = 15$.

b. $8y - 4x < 68, (3.1, 10)$

$8y - 4x < 68$	Write original equation.
$8(10) - 4(3.1) < 68$	Substitute 3.1 for x and 10 for y.
$80 - 12.4 < 68$	Simplify.
$67.6 < 68$	This is a true statement.

$(3.1, 10)$ is a solution of the equation $8y - 4x < 68$.

Exercises for Example 3

Decide whether the given ordered pair is a solution of the equation or inequality.

9. $5x + y = 8, (3, -7)$

10. $-12x = 8y - 1, (2.5, 9.6)$

11. $\frac{1}{2}x - y > 2 + 4x, (-3, -9)$

12. $3x - 4.6y \leq 16, (4, -6)$

NAME _______________ DATE _______________

Strategies for Reading Mathematics

For use with Chapter 7

Strategy: Reading for Information

One of the most important problem-solving skills is figuring out how to use what you know about a problem to solve for the unknowns. Ask yourself, "What is the question? What do I know? What do I need to find out?" Be careful. Sometimes you need to solve for more than one unknown. The problem below is from Example 3 on page 407.

> **Museum Admissions** In one day the National Civil Rights Museum in Memphis Tennessee, collected $1590 from 321 people admitted to the museum. The price of each adult admission is $6. People with the ages of 4–17 may pay the child admission, $4. Estimate how many adults and how many children were admitted that day.

Don't be distracted by unnecessary information. You don't need to know the children's ages.

Determine the unknowns and assign variables.
Number of adults $= x$
Number of children $= y$

Determine what you know.
Total collected = $1590
Total number of people = 321
Adult admission = $6
Child admission = $4

Read each problem carefully and pay attention to facts, ideas, and special word meanings. For example, in the problem above, the total amount collected is given. The word *total* suggests addition. Use that to help you set up a verbal, algebraic, and/or graphic representation. Remember, reread as often as necessary.

STUDY TIP
Explain Your Solution
One way to check your solution to a problem is to explain your process out loud to a partner or to yourself. It may help you catch errors.

STUDY TIP
Choose an Appropriate Strategy
By now you know many strategies. Choose the one that works best for your problem. Then check that your answer makes sense.

Questions

Use the following problem to answer questions 1–4.

In a class of 34 students, 24 students play a fall sport. Two-thirds of the boys and three-quarters of the girls play a fall sport. How many girls and how many boys are in the class?

1. Label the knowns and unknowns in the problem.
2. Do you have enough information to write equations to solve this problem? Explain your thinking.
3. Why do you need two equations to solve this problem? Do you always need two equations to solve a problem? Explain your reasoning.
4. Suppose you solved the problem and ended up with the answer $x = -18$. Explain why your answer is or is not reasonable.

CHAPTER 7 CONTINUED

NAME ______________________ DATE __________

Strategies for Reading Mathematics

For use with Chapter 7

Visual Glossary

The Study Guide on page 396 lists the key vocabulary for Chapter 7 as well as review vocabulary from previous chapters. Use the page references on page 396 or the Glossary in the textbook to review key terms from prior chapters. Use the visual glossary below to help you understand some of the key vocabulary in Chapter 7. You may want to copy these diagrams into your notebook and refer to them as you complete the chapter.

GLOSSARY

linear system (p. 398) Two or more linear equations in the same variables. This is also called a system of linear equations.

linear combination (p. 411) An equation obtained by adding one of two equations (or a multiple of one of the equations) to the other equation in a linear system.

solution of a linear system (p. 398) An ordered pair (x, y) that satisfies each equation in the system.

system of linear inequalities (p. 432) Two or more linear inequalities in the same variables. This is also called a system of inequalities.

graph of a system of linear inequalities (p. 432) The graph of all solutions of the system.

solution of a system of linear inequalities (p. 432) An ordered pair (x, y) that is a solution of each inequality in the system.

Solution of a Linear System

The two equations $3x + y = 6$ and $4x - y = 8$ together form a linear system of equations. The algebraic solution and the graph below both show the solution of the linear system.

linear system $\begin{cases} 3x + y = 6 \\ 4x - y = 8 \end{cases}$

$$\begin{array}{r} 3x + y = 6 \\ +\ 4x - y = 8 \\ \hline 7x \quad = 14 \\ x = \mathbf{2} \end{array}$$

linear combination

$3(2) + y = 6$

$6 + y = 6$

$y = \mathbf{0}$

(2, 0) is the solution of the linear system.

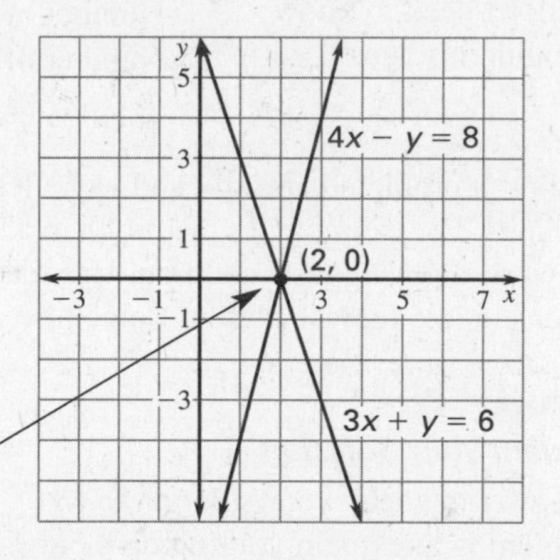

The Graph of a System of Linear Inequalities

The graph of a linear inequality is indicated by a shaded half-plane. When you graph a system of linear inequalities, the graph of the system is indicated by the most heavily shaded region of the plane.

The graph is the graph of the following system:

$x > -4$

$y \le 4$

$y \ge x + 4$

graph of the system (the dark triangle)

$(-2, 3)$ is a solution:

$-2 > -4, \quad 3 \le 4, \quad 3 \ge -2 + 4$

LESSON 7.1

TEACHER'S NAME ____________________ CLASS ______ ROOM ______ DATE ______

Lesson Plan

1-day lesson (See *Pacing the Chapter,* TE pages 394C–394D)

For use with pages 397–404

GOALS
1. **Solve a system of linear inequalities by graphing.**
2. **Use a system of linear inequalities to model a real-life situation.**

State/Local Objectives ____

____

✓ Check the items you wish to use for this lesson.

STARTING OPTIONS

____ Prerequisite Skills Review: CRB pages 5–6
____ Strategies for Reading Mathematics: CRB pages 7–8
____ Warm-Up or Daily Homework Quiz: TE pages 398 and 381, CRB page 11, or Transparencies

TEACHING OPTIONS

____ Motivating the Lesson: TE page 399
____ Concept Activity: SE page 397; CRB page 12 (Activity Support Master)
____ Lesson Opener (Visual Approach): CRB page 13 or Transparencies
____ Graphing Calculator Activity with Keystrokes: CRB page 14
____ Examples 1–3: SE pages 398–400
____ Extra Examples: TE pages 399–400 or Transparencies; Internet
____ Technology Activity: SE page 404
____ Closure Question: TE page 400
____ Guided Practice Exercises: SE page 401

APPLY/HOMEWORK

Homework Assignment

____ Basic 12–34 even, 35, 41, 42, 50, 52, 56, 59
____ Average 12–34 even, 35, 36, 41, 42, 50, 52, 56, 59
____ Advanced 12–34 even, 36–46, 50, 52, 56, 59

Reteaching the Lesson

____ Practice Masters: CRB pages 15–17 (Level A, Level B, Level C)
____ Reteaching with Practice: CRB pages 18–19 or Practice Workbook with Examples
____ Personal Student Tutor

Extending the Lesson

____ Applications (Real-Life): CRB page 21
____ Challenge: SE page 403; CRB page 22 or Internet

ASSESSMENT OPTIONS

____ Checkpoint Exercises: TE pages 399–400 or Transparencies
____ Daily Homework Quiz (7.1): TE page 403, CRB page 25, or Transparencies
____ Standardized Test Practice: SE page 403; TE page 403; STP Workbook; Transparencies

Notes ____

____

____

LESSON 7.1

TEACHER'S NAME ______ CLASS ______ ROOM ______ DATE ______

Lesson Plan for Block Scheduling

Half-day lesson (See *Pacing the Chapter,* TE pages 394C–394D) **For use with pages 397–404**

GOALS
1. **Solve a system of linear inequalities by graphing.**
2. **Use a system of linear inequalities to model a real-life situation.**

State/Local Objectives ______

CHAPTER PACING GUIDE	
Day	**Lesson**
1	**7.1 (all)**; 7.2 (all)
2	7.3 (all)
3	7.4 (all)
4	7.5 (all); 7.6 (begin)
5	7.6 (end); Review Ch. 7
6	Assess Ch. 7; 8.1 (begin)

✓ Check the items you wish to use for this lesson.

STARTING OPTIONS

____ Prerequisite Skills Review: CRB pages 5–6
____ Strategies for Reading Mathematics: CRB pages 7–8
____ Warm-Up or Daily Homework Quiz: TE pages 398 and 381, CRB page 11, or Transparencies

TEACHING OPTIONS

____ Motivating the Lesson: TE page 399
____ Concept Activity: SE page 397; CRB page 12 (Activity Support Master)
____ Lesson Opener (Visual Approach): CRB page 13 or Transparencies
____ Graphing Calculator Activity with Keystrokes: CRB page 14
____ Examples 1–3: SE pages 398–400
____ Extra Examples: TE pages 399–400 or Transparencies; Internet
____ Technology Activity: SE page 404
____ Closure Question: TE page 400
____ Guided Practice Exercises: SE page 401

APPLY/HOMEWORK

Homework Assignment (See also the assignment for Lesson 7.2.)
____ Block Schedule: 12–34 even, 35, 36, 41, 42, 50, 52, 56, 59

Reteaching the Lesson
____ Practice Masters: CRB pages 15–17 (Level A, Level B, Level C)
____ Reteaching with Practice: CRB pages 18–19 or Practice Workbook with Examples
____ Personal Student Tutor

Extending the Lesson
____ Applications (Real-Life): CRB page 21
____ Challenge: SE page 403; CRB page 22 or Internet

ASSESSMENT OPTIONS

____ Checkpoint Exercises: TE pages 399–400 or Transparencies
____ Daily Homework Quiz (7.1): TE page 403, CRB page 25, or Transparencies
____ Standardized Test Practice: SE page 403; TE page 403; STP Workbook; Transparencies

Notes ______

NAME ______________________ DATE __________

Available as a transparency

WARM-UP EXERCISES

For use before Lesson 7.1, pages 397–404

Is the ordered pair a solution of the equation $2x - 3y = 5$?

1. (1, 0)

2. (−1, 1)

3. (1, −1)

4. (4, 1)

DAILY HOMEWORK QUIZ

For use after Lesson 6.7, pages 375–382

Make a box-and-whisker plot of the data.

1. Waiting times at a doctor's office (min)

22 45 52 37 43 11 62 29
39 28 44 33 28 41 39 28

2. Highway fuel efficiencies (mi/gal)

21 18 26 31 25 33 41 20 18
19 24 28 34 55 37 28 30 25

3. In the plot from Question 2 above, compare the spread of the data in the lowest fourth and highest fourth of the data values.

NAME _______________ DATE _______

Activity Support Master

For use with page 397

Ordered pair that you represent: _______________

Equation 1: $x - y = -2$

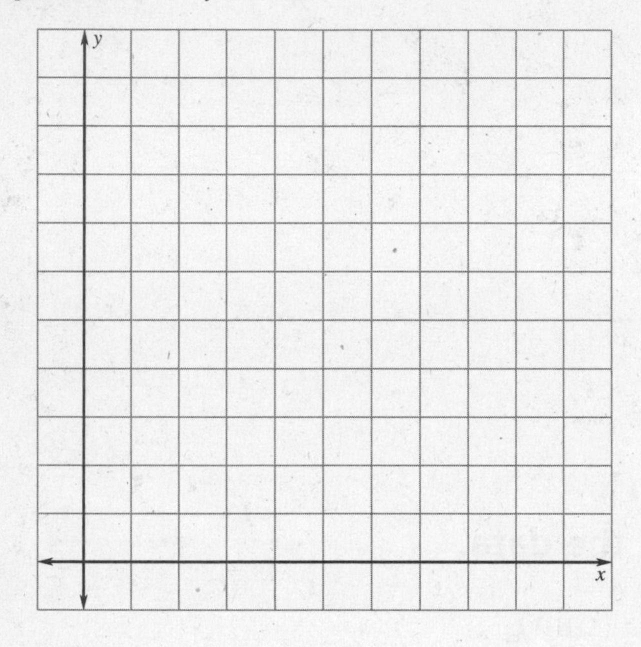

Equation 2: $x + y = 4$

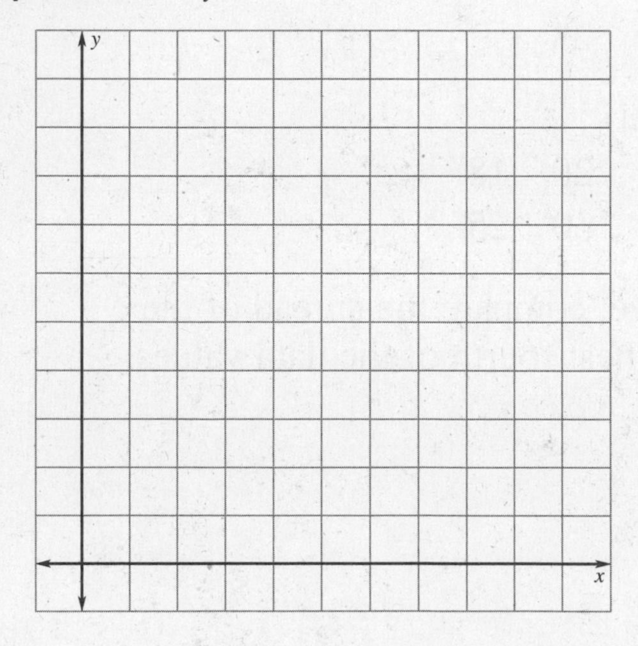

NAME ______________________ DATE __________

Available as a transparency

Visual Approach Lesson Opener

For use with pages 398–403

The graph of the equations x + y = 2 and x − y = 4 is shown at the right.

1. What are the coordinates of the point of intersection?

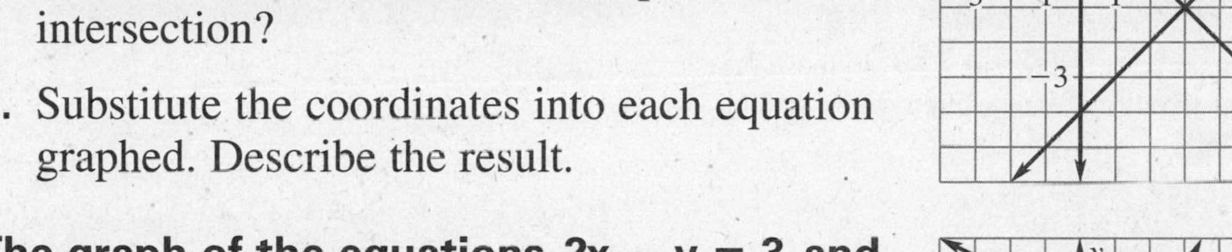

2. Substitute the coordinates into each equation graphed. Describe the result.

The graph of the equations 2x − y = 3 and x + 2y = 4 is shown at the right.

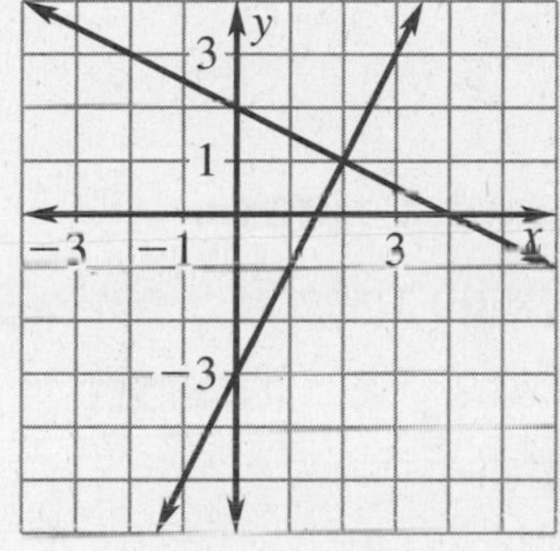

3. What are the coordinates of the point of intersection?

4. Substitute the coordinates into each equation graphed. Describe the result.

The graph of the equations x − y = 2 and −3x + y = 2 is shown at the right.

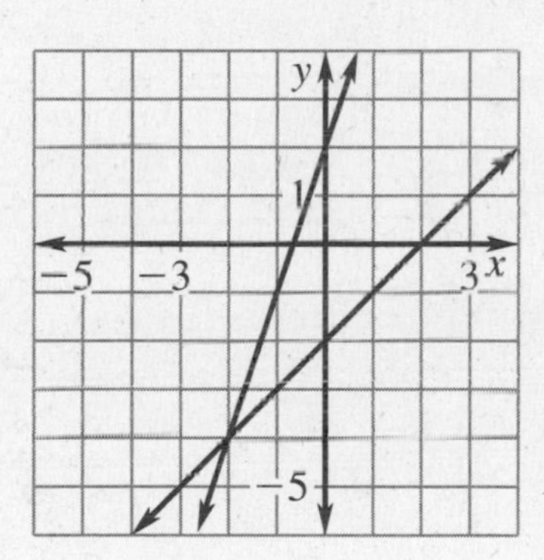

5. What are the coordinates of the point of intersection?

6. Substitute the coordinates into each equation graphed. Describe the result.

7. Make a conjecture about the coordinates of the point of intersection of two linear equations.

LESSON 7.1

NAME ______________________ DATE __________

Graphing Calculator Activity Keystrokes

For use with Technology Activity 7.1 on page 404

TI-82

Y= (-) 0.3 X,T,θ + 1.8 ENTER

0.6 X,T,θ − 1.5 ENTER

ZOOM 6 2nd [CALC] 5 ENTER ENTER

Use the cursor keys, ◀ and ▶, to move the trace cursor to point of intersection near $x = 3.6$. Press ENTER.

TI-83

Y= (-) 0.3 X,T,θ,n + 1.8 ENTER

0.6 X,T,θ,n − 1.5 ENTER

ZOOM 6 2nd [CALC] 5 ENTER ENTER

3.6 ENTER

SHARP EL-9600c

Y= (-) 0.3 X/θ/T/n + 1.8 ENTER

0.6 X/θ/T/n − 1.5 ENTER

ZOOM [A] 5

2ndF [CALC] 2

CASIO CFX-9850GA PLUS

From the main menu, choose GRAPH.

(-) 0.3 X,θ,T + 1.8 EXE

0.6 X,θ,T − 1.5 EXE

SHIFT F3 F3 EXIT

F6 F5 F5

Lesson 7.1

LESSON 7.1

NAME ______________________________ DATE ____________

Practice A

For use with pages 398–403

Decide whether the ordered pair is a solution of the system of linear equations.

1. $(1, 1), (-1, 0)$
$2x + y = 3$
$x - 2y = -1$

2. $(-2, 4), (3, -4)$
$4x + y = -4$
$-x - y = 1$

3. $(5, 4), (4, 1)$
$x - y = 3$
$3x - y = 11$

4. $(-6, -4), (3, -1)$
$x - 3y = 6$
$2x - y = -8$

5. $(-3, -4), (-1, 4)$
$-4x + y = 8$
$5x - 3y = -3$

6. $(0, -2), (-6, 2)$
$-2x - 3y = 6$
$3x + 4y = -10$

Use the graph to solve the linear system. Check your solution algebraically.

7. $-x + y = 4$
$x + y = 4$

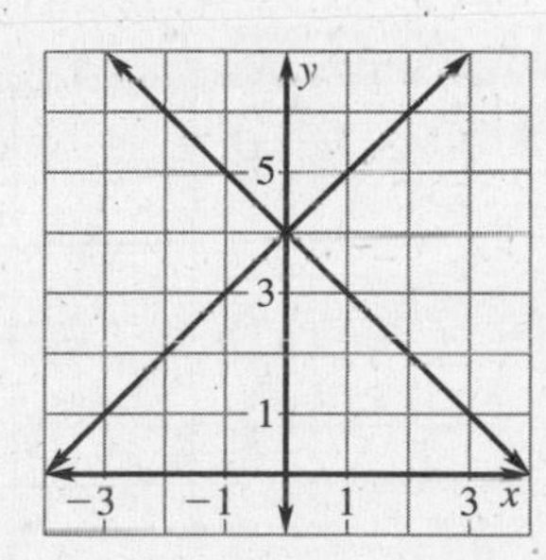

8. $x + y = 0$
$-x + y = -2$

9. $-x + y = 2$
$2x + y = 8$

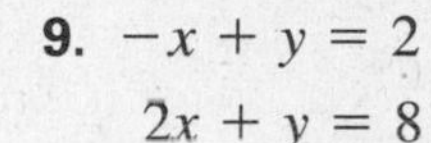

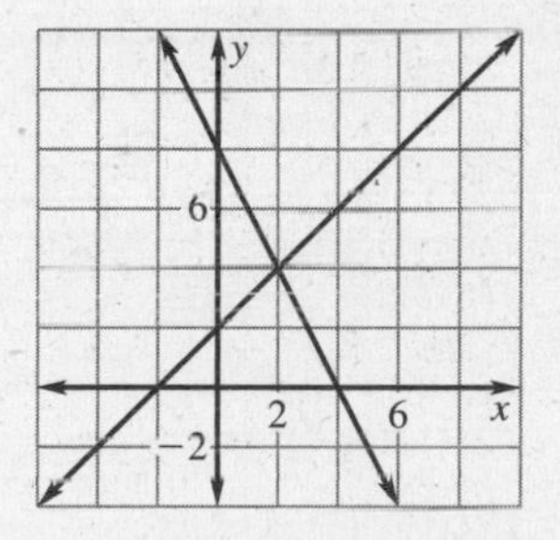

Graph and check to solve the linear system.

10. $x = 6$
$y = -2$

11. $y = x - 2$
$y = -x + 4$

12. $y = -2x - 4$
$y = -\frac{1}{2}x - 1$

13. $3x + y = 6$
$-x + y = -2$

14. $-2x + y = 1$
$y = 5$

15. $x + 2y = 6$
$-3x + y = 10$

16. ***Juice*** You bought 12 1-gallon bottles of apple and orange juice for a school dance. The apple juice was on sale for $1.00 per gallon bottle. The orange juice was $1.50 per 1 gallon bottle. You spent $15.00. Assign labels to the verbal model below. Write an algebraic model. How many bottles of each type of juice did you buy?

[Number of bottles of apple juice] + [Number of bottles of orange juice] = [Total number of bottles]

[Price per apple juice bottle] · [Number of bottles of apple juice] + [Price per orange juice bottle] · [Number of bottles of orange juice] = [Total price]

NAME ______________________ DATE __________

Practice B

For use with pages 398–403

Decide whether the ordered pair is a solution of the system of linear equations.

1. (1, 1), (0, 3)
$2x + y = 3$
$x - 2y = -1$

2. (2, 4), (−3, 8)
$4x + y = -4$
$-x - y = 1$

3. (−5, −2), (4, 1)
$x - y = 3$
$3x - y = 11$

4. (−6, −4), (−4, 0)
$x - 3y = 6$
$2x - y = -8$

5. (−3, −4), (3, 6)
$-4x + y = 8$
$5x - 3y = -3$

6. (3, −4), (−6, 2)
$-2x - y = 6$
$3x + 4y = -10$

Use the graph to solve the linear system. Check your solution algebraically.

7. $-x + y = -8$
$x + y = 4$

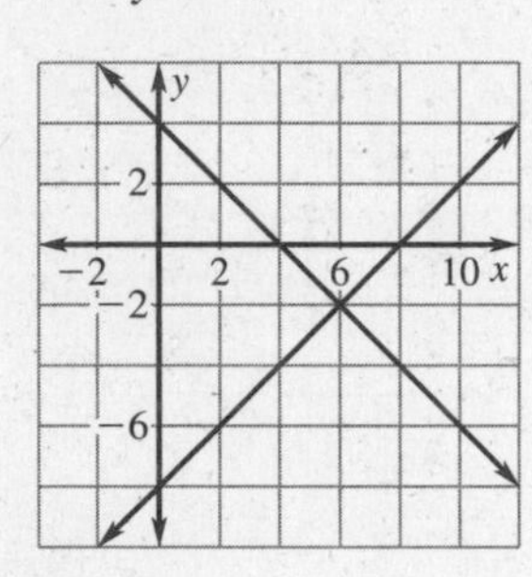

8. $3x + y = -6$
$-x - 2y = -3$

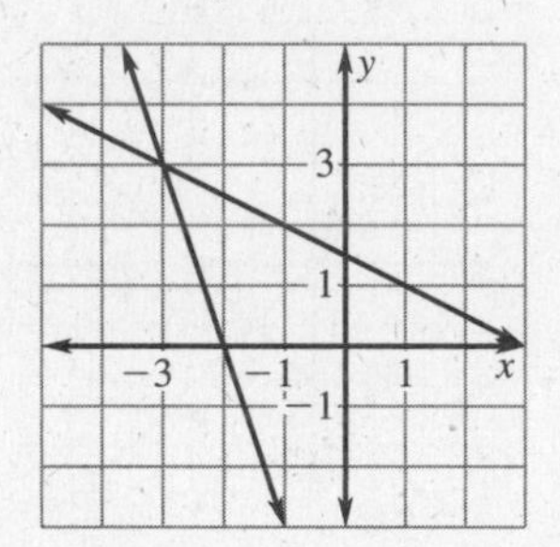

9. $4x + 2y = -12$
$2x + 2y = 8$

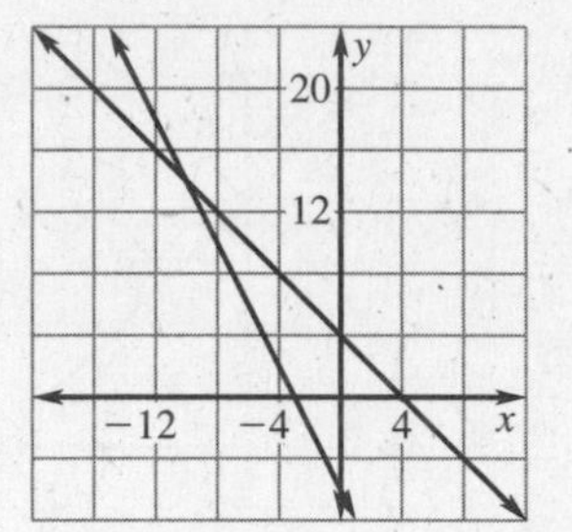

Graph and check to solve the linear system.

10. $x = 6$
$y = -3$

11. $y = x - 2$
$y = -x - 4$

12. $y = 2x - 4$
$y = -\frac{1}{2}x + 1$

13. $-3x + y = 6$
$-x + y = -2$

14. $x + 2y = -6$
$-3x + y = -10$

15. $y = \frac{1}{2}x + 3$
$y = x + 4$

16. ***Juice*** You bought 12 1-gallon bottles of apple and orange juice for a school dance. The apple juice was on sale for \$1.00 per gallon bottle. The orange juice was \$1.75 per 1-gallon bottle. You spent \$15.00. Assign labels to the verbal model below. Write an algebraic model. How many bottles of each type of juice did you buy?

[Number of bottles of apple juice] + [Number of bottles of orange juice] = [Total number of bottles]

[Price per apple juice bottle] · [Number of bottles of apple juice] + [Price per orange juice bottle] · [Number of bottles of orange juice] = [Total price]

17. ***Baseball Outs*** In a game, 18 of a baseball team's 27 outs were fly balls. Fifty percent of the outs made by infielders and 100% of the outs made by outfielders were fly balls. How many outs were made by infielders? How many outs were made by outfielders? (Hint: Write one equation for the total number of outs and another equation for the number of fly ball outs.)

NAME ______________________ DATE __________

Practice C

For use with pages 398–403

Decide whether the ordered pair is a solution of the system of linear equations.

1. $(1, 1), (-1, 5)$
$2x + y = 3$
$x - 2y = -1$

2. $(-2, 4), (-1, 0)$
$4x + y = -4$
$-x - y = 1$

3. $(-5, -8), (4, 1)$
$x - y = 3$
$3x - y = 11$

4. $(-6, -4), (0, -8)$
$x - 3y = 6$
$2x - y = -8$

5. $(-6, -9), (-4, 8)$
$-4x + y = 8$
$5x - 3y = -3$

6. $(6, -7), (-6, 2)$
$-2x - 3y = 6$
$3x + 4y = -10$

Use the graph to solve the linear system. Check your solution algebraically.

7. $-x + y = -8$
$2x + y = 4$

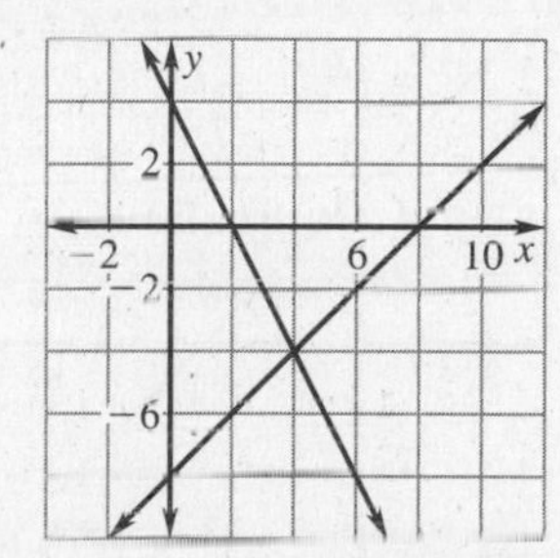

8. $3x + y = 6$
$-x + y = -4$

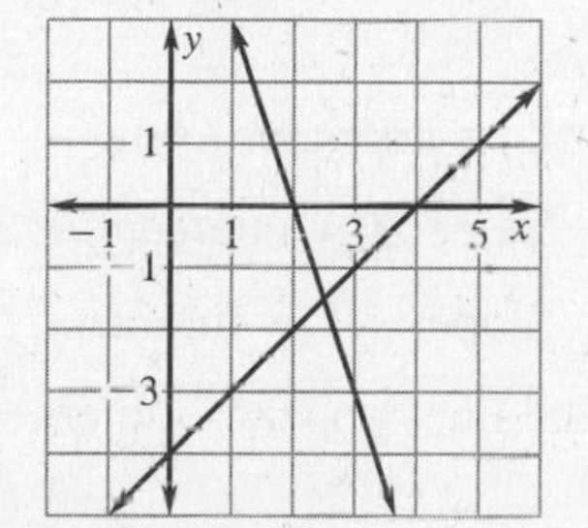

9. $4x + 2y = 12$
$2x - 3y = 10$

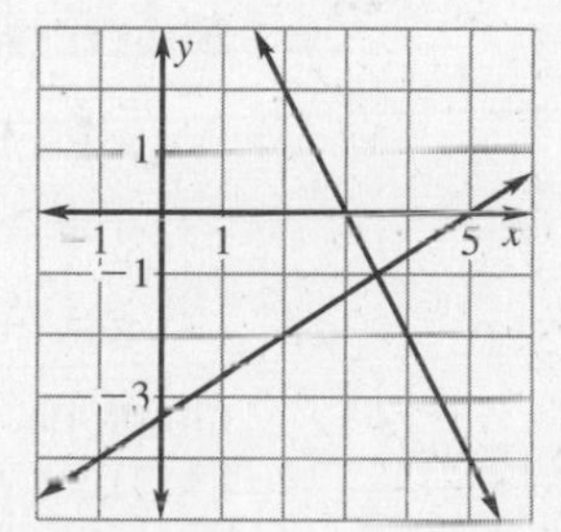

Graph and check to solve the linear system.

10. $-3x + y = 8$
$-x + y = -2$

11. $-2x + y = 1$
$y = -5$

12. $x - 2y = 7$
$-5x + y = 10$

13. $y = -4x - 2$
$y = -2x + 1$

14. $y = \frac{1}{2}x + 9$
$y = -x + 6$

15. $3x - 5y = -30$
$x - 5y = -20$

16. ***Buying Juice*** You bought 12 bottles of apple juice and orange juice. The apple juice was on sale for \$1.00 per bottle. The orange juice was \$1.75 per bottle. You spent \$15.00. How many bottles of each type of juice did you buy? (Hint: Write one equation for the total number of bottles and another equation for the total price.)

17. ***Investments*** A total of \$25,000 is invested in two funds paying 5% and 6% annual interest. The combined annual interest is \$1400. How much of the \$25,000 is invested in each type of fund? (Hint: Write one equation for the amount invested in each fund and another for the interest earned.)

18. ***Umbrella Sales*** The matrix gives the number of automatic and manual opening umbrellas sold at a shop in 1985 and 1995. Use a linear model to represent the sales of each type of umbrella. Let $t = 0$ correspond to 1985. Sketch the graphs and estimate when the number of automatic umbrellas sold equaled the number of manual umbrellas sold.

	1985	1995
Automatic	10	20
Manual	25	10

NAME ______________________ DATE ________

Reteaching with Practice

For use with pages 398–403

GOAL **Solve a system of linear equations by graphing and model a real-life problem using a linear system**

VOCABULARY

Two equations in two variables form a **system of linear equations** or simply a **linear system.**

A **solution of a system of linear equations** in two variables is an ordered pair (x, y) that satisfies each equation in the system.

EXAMPLE 1 *Using the Graph-and-Check Method*

Solve the linear system graphically. Check the solution algebraically.

$-3x + y = -7$ Equation 1

$2x + 2y = 10$ Equation 2

SOLUTION

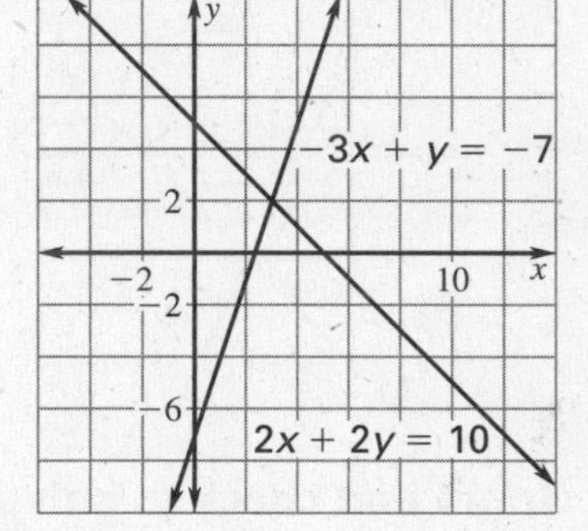

Write each equation in slope-intercept form.

$y = 3x - 7$ Slope: 3, y-intercept: -7

$y = -x + 5$ Slope: -1, y-intercept: 5

Graph each equation. The lines appear to intersect at (3, 2).

To check (3, 2) as a solution algebraically, substitute 3 for x and 2 for y in each original equation.

EQUATION 1	EQUATION 2
$-3x + y = -7$	$2x + 2y = 10$
$-3(3) + 2 \stackrel{?}{=} -7$	$2(3) + 2(2) \stackrel{?}{=} 10$
$-7 = -7$	$10 = 10$

Because (3, 2) is a solution of each equation in the linear system, it is a solution of the linear system.

Exercises for Example 1

Graph and check to solve each linear system.

1. $y = -x + 5$
$y = x + 1$

2. $2x - y = 2$
$x = 4$

3. $2x + y = 2$
$x - y = 4$

NAME ______________________ DATE __________

Reteaching with Practice

For use with pages 398–403

EXAMPLE 2 Using a Linear System to Model a Real-Life Problem

Tickets for the theater are $5 for the balcony and $10 for the orchestra. If 600 tickets were sold and the total receipts were $4750, how many tickets were sold for the orchestra?

SOLUTION

Verbal Model

[Number of balcony tickets] + [Number of orchestra tickets] = [Total number of tickets]

[Price of balcony tickets] · [Number of balcony tickets] + [Price of orchestra tickets] · [Number of orchestra tickets] = [Total receipts]

Labels

Price of balcony tickets = 5 (dollars)

Number of balcony tickets = x (tickets)

Price of orchestra tickets = 10 (dollars)

Number of orchestra tickets = y (tickets)

Total number of tickets = 600 (tickets)

Total receipts = 4750 (dollars)

Algebraic Model

$x + y = 600$ Equation 1 (tickets)

$5x + 10y = 4750$ Equation 2 (receipts)

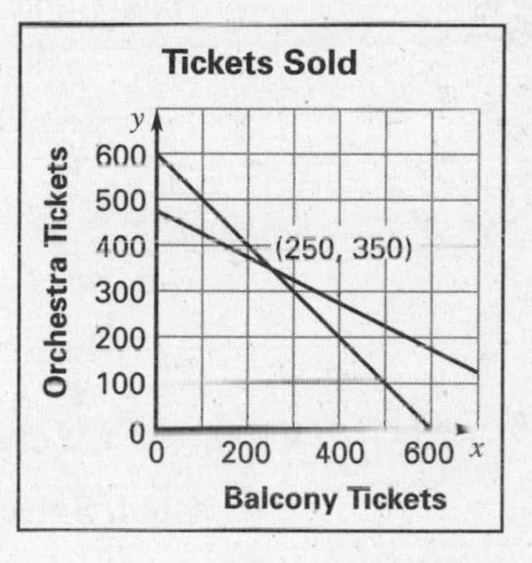

Graph the system.
Check the solution:

$$250 + 350 = 600 \text{ and } 5(250) + 10(350) = 1250 + 3500 = 4750.$$

350 orchestra tickets were sold.

Exercises for Example 2

4. Rework Example 2 if 800 tickets were sold.

5. Rework Example 2 if total receipts were $3500.

NAME ______________________________ DATE ____________

Quick Catch-Up for Absent Students

For use with pages 397–404

The items checked below were covered in class on (date missed) ____________

Activity 7.1: Investigating Graphs of Linear Systems (p. 397)

____ **Goal:** Determine if two different linear equations have a solution in common.

____ Student Help: Look Back

Lesson 7.1: Solving Linear Systems by Graphing

____ **Goal 1:** Solve a system of linear equations by graphing. (pp. 398–399)

Material Covered:

____ Student Help: Look Back

____ Example 1: Checking the Intersection Point

____ Example 2: Using the Graph-and-Check Method

Vocabulary:

system of linear equations, p. 398 linear system, p. 398

solution of a system of linear equations, p. 398

____ **Goal 2:** Model a real-life problem using a linear system. (p. 400)

Material Covered:

____ Example 3: Writing and Using a Linear System

Activity 7.1: Solving Linear Systems by Graphing (p. 404)

____ **Goal:** Solve a linear system using a graphing calculator.

____ Student Help: Keystroke Help

____ Other (specify) ______________________________

____

Homework and Additional Learning Support

____ Textbook (specify) pp. 401–403

____ Internet: Extra Examples at www.mcdougallittell.com

____ *Reteaching with Practice* worksheet (specify exercises) ______________

____ *Personal Student Tutor* for Lesson 7.1

NAME ______________________________ DATE __________

Real-Life Application: When Will I Ever Use This?

For use with pages 398–403

Newspaper Routes

Newspapers are publications devoted chiefly to presenting and commenting on the news. Newspapers provide an excellent means of keeping well informed on current events. They also play a vital role in shaping public opinion.

Newspapers have certain advantages over the other major news media. They can cover more news and in much greater detail, than can television and radio newscasts. Newspapers are the major source of local news. The United States has about 1500 daily newspapers and 8000 weekly and semiweekly newspapers. In 1999, the total circulation of daily papers in the United States was about 56 million and the total Sunday circulation was about 60 million.

In Exercises 1-5, use the following information.

You are about to start working for the newspaper publisher. You are offered two routes to deliver papers. Route 1 has 36 current customers and has about 3 new subscriptions a month. Route 2 is a newer route, and only has 20 current customers, but is gaining about 5 new subscriptions per month.

1. For each newspaper route, write a linear model to represent the total number of subscriptions.
2. Graph the linear equations.
3. Using your graph, estimate when the two routes will have the same number of customers.
4. Using your graph, determine which route will have more customers after 6 months? After 12 months?
5. You plan on delivering papers for at least a year. The more newspapers you deliver, the more money you make. Which route would you choose to deliver newspapers?

In Exercises 6 and 7, use the following information.

Two years after your first day of delivering newspapers, the publisher decides to split your route in half.

6. Using your graph, find how many newspapers you will be delivering after two years.
7. How many newspapers will you be delivering after your route is split in half?

NAME ______________________ DATE __________

Challenge: Skills and Applications

For use with pages 398–403

In Exercises 1–2, decide whether the ordered pair is a solution of the system of linear equations.

1. $8x - 6y = 2$ $\left(\frac{1}{2}, \frac{1}{3}\right)$
   $7x + 2y = 4\frac{1}{2}$

2. $4x - 5y = \frac{2}{3}$ $\left(\frac{2}{3}, \frac{2}{5}\right)$
   $9x - 4y = 4\frac{2}{5}$

In Exercises 3–4, use the table below, which gives the numbers of users of two Internet providers in a small town.

	1995	*2000*
Provider A	345	580
Provider B	273	628

3. For each provider, write a linear model to represent the number of users at time t, where t represents the number of years since 1995.

4. Use a graph to estimate when the two providers had the same number of users.

In Exercises 5–7, use the information in the table, which gives the population of three cities based on July 1994 estimates and gives the growth rate of each city.

City	*Population*	*Growth rate (people per year)*
City A	547,725	−25,195
City B	493,559	27,146
City C	237,612	12,831

5. For each city, write a linear model to represent the population of the city at time t, where t represents the number of years since 1994.

6. Use a graph to estimate when City A and City B should have the same population.

7. Use a graph to estimate when City A and City C should have the same population.

LESSON 7.2

TEACHER'S NAME ____________________ CLASS ______ ROOM ______ DATE ______

Lesson Plan

1-day lesson (See *Pacing the Chapter,* TE pages 394C–394D) For use with pages 405–410

GOALS
1. **Use substitution to solve a linear system.**
2. **Model a real-life situation using a linear system.**

State/Local Objectives ____________________

____

✓ **Check the items you wish to use for this lesson.**

STARTING OPTIONS

____ Homework Check: TE page 401; Answer Transparencies
____ Warm-Up or Daily Homework Quiz: TE pages 405 and 403, CRB page 25, or Transparencies

TEACHING OPTIONS

____ Motivating the Lesson: TE page 406
____ Lesson Opener (Activity): CRB page 26 or Transparencies
____ Graphing Calculator Activity with Keystrokes: CRB page 27
____ Examples 1–3: SE pages 405–407
____ Extra Examples: TE pages 406–407 or Transparencies; Internet
____ Closure Question: TE page 407
____ Guided Practice Exercises: SE page 408

APPLY/HOMEWORK

Homework Assignment

____ Basic 14–44 even, 35, 48–51, 55, 60, 65, 68
____ Average 14–44 even, 35, 43, 48–51, 55, 60, 65, 68
____ Advanced 14–44 even, 35, 43, 48–53, 55, 60, 65, 68

Reteaching the Lesson

____ Practice Masters: CRB pages 28–30 (Level A, Level B, Level C)
____ Reteaching with Practice: CRB pages 31–32 or Practice Workbook with Examples
____ Personal Student Tutor

Extending the Lesson

____ Applications (Interdisciplinary): CRB page 34
____ Challenge: SE page 410; CRB page 35 or Internet

ASSESSMENT OPTIONS

____ Checkpoint Exercises: TE pages 406–407 or Transparencies
____ Daily Homework Quiz (7.2): TE page 410, CRB page 38, or Transparencies
____ Standardized Test Practice: SE page 410; TE page 410; STP Workbook; Transparencies

Notes ____________________

____

____

LESSON 7.2

TEACHER'S NAME ______________________ CLASS ______ ROOM ______ DATE ______

Lesson Plan for Block Scheduling

Half-day lesson (See *Pacing the Chapter,* TE pages 394C–394D) **For use with pages 405–410**

GOALS
1. **Use substitution to solve a linear system.**
2. **Model a real-life situation using a linear system.**

State/Local Objectives ______________________________

____

____

CHAPTER PACING GUIDE	
Day	**Lesson**
1	7.1 (all); **7.2 (all)**
2	7.3 (all)
3	7.4 (all)
4	7.5 (all); 7.6 (begin)
5	7.6 (end); Review Ch. 7
6	Assess Ch. 7; 8.1 (begin)

✓ **Check the items you wish to use for this lesson.**

STARTING OPTIONS

____ Homework Check: TE page 401; Answer Transparencies
____ Warm-Up or Daily Homework Quiz: TE pages 405 and 403, CRB page 25, or Transparencies

TEACHING OPTIONS

____ Motivating the Lesson: TE page 406
____ Lesson Opener (Activity): CRB page 26 or Transparencies
____ Graphing Calculator Activity with Keystrokes: CRB page 27
____ Examples 1–3: SE pages 405–407
____ Extra Examples: TE pages 406–407 or Transparencies; Internet
____ Closure Question: TE page 407
____ Guided Practice Exercises: SE page 408

APPLY/HOMEWORK

Homework Assignment (See also the assignment for Lesson 7.1.)

____ Block Schedule: 14–44 even, 35, 43, 48–51, 55, 60, 65, 68

Reteaching the Lesson

____ Practice Masters: CRB pages 28–30 (Level A, Level B, Level C)
____ Reteaching with Practice: CRB pages 31–32 or Practice Workbook with Examples
____ Personal Student Tutor

Extending the Lesson

____ Applications (Interdisciplinary): CRB page 34
____ Challenge: SE page 410; CRB page 35 or Internet

ASSESSMENT OPTIONS

____ Checkpoint Exercises: TE pages 406–407 or Transparencies
____ Daily Homework Quiz (7.2): TE page 410, CRB page 38, or Transparencies
____ Standardized Test Practice: SE page 410; TE page 410; STP Workbook; Transparencies

Notes ______________________________

____

____

LESSON **7.2**

NAME ____________________ DATE ____________

Available as a transparency

WARM-UP EXERCISES

For use before Lesson 7.2, pages 405–410

Is $(-3, 2)$ a solution of the equation?

1. $3x - y = 4$

2. $x + 2y = 1$

3. Is $(4, 0)$ a solution of the system of equations?

$2x + y = 8$

$-x + 3y = 4$

Solve each equation for *y*.

4. $2x + y = 12$

5. $-3x - 2y = -2$

DAILY HOMEWORK QUIZ

For use after Lesson 7.1, pages 397–404

1. Decide whether $(-3, 5)$ is a solution of the linear system.

$2x - 5y = -31; \quad -3x + y = 14$

2. Use the graph to solve the linear system.

$y = 2x + 3; \quad 3x + 2y = -1$

Graph and check to solve the linear system.

3. $2x + 4y = -2$

$-x + 2y = -7$

4. $\frac{3}{7}x + \frac{1}{7}y = -\frac{5}{7}$

$-\frac{2}{7}x + \frac{3}{7}y = -\frac{4}{7}$

5. The Smith family made an \$800 downpayment and pays \$75 a month for new furniture. At the same time, the Cooper family made a \$500 downpayment and pays \$95 a month for their new furniture. Use a graph to determine how many months it will be before the amounts they have paid are equal.

LESSON 7.2

NAME ______________________ DATE ____________

Available as a transparency

Activity Lesson Opener

For use with pages 405–410

SET UP: Work in a group. YOU WILL NEED • index cards.

1. Write one system of equations on each index card. Shuffle the cards and place them face down on the desk in 3 rows of 3.

1. $y = x + 2$ $y + x = 4$	2. $x + 2y = 4$ $x = y - 1$	3. $3x + y = 4$ $y = x + 5$
4. $x = y$ $x - 3y = 3$	5. $y + 2x = 6$ $y = 2 - x$	6. $x - y = 7$ $x = 4y$
7. $4x + y = 9$ $y = 3x$	8. $5y + x = 6$ $x = y + 3$	9. $x - y = 7$ $y = 3x$

2. On the second group of index cards, write one equation on each card. Place these cards face up on the desk.

a. $4y - y = 7$	b. $5y + (y + 3) = 6$	c. $(2 - x) + 2x = 6$
d. $4x + 3x = 9$	e. $3x + (x + 5) = 4$	f. $(y - 1) + 2y = 4$
g. $x - 3x = 7$	h. $(x + 2) + x = 4$	i. $y - 3y = 3$

3. The first group chooses one of the equation cards from the face-up cards. Then, they turn over 3 system of equation cards. If the equation can be found using one of the systems chosen, someone in the group picks up the two cards. If not, the cards are returned to their appropriate positions, and the next group gets a turn. For example, in the system $x = 2y$, $x + 2y = 5$, the equation $(2y) + 3y = 5$ is found by replacing x in the second equation with its equivalent value, $2y$, from the first equation in the system. This system and equation are "a match." The game is over when all cards have been matched. The group with the most cards at the end of the game "wins."

4. When the game is over, look at the pairs of matched cards. How was the equation in each pair found using the system of equations?

Lesson 7.2

LESSON 7.2

NAME ______________________ DATE __________

Graphing Calculator Activity Keystrokes

For use with page 409

Keystrokes for Exercise 36

TI-82

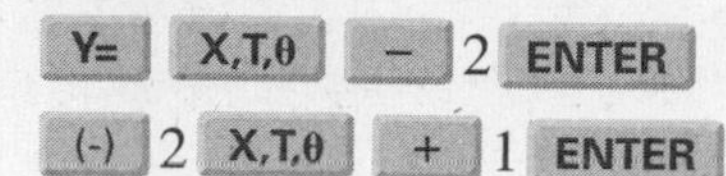

Y= X,T,θ − 2 ENTER

(-) 2 X,T,θ + 1 ENTER

ZOOM 6 2nd [CALC] 5 ENTER ENTER

Use the cursor keys, ◀ and ▶, to move the trace cursor to point of intersection near $x =$ 1. Press ENTER .

TI-83

Y= X,T,θ,*n* − 2 ENTER

(-) 2 X,T,θ,*n* + 1 ENTER

ZOOM 6 2nd [CALC] 5 ENTER ENTER 1

ENTER

SHARP EL-9600C

Y= X/θ/T/*n* − 2 ENTER

(-) 2 X/θ/T/*n* + 1 ENTER

ZOOM [A] 5

2ndF [CALC] 2

CASIO CFX-9850GA PLUS

From the main menu, choose GRAPH.

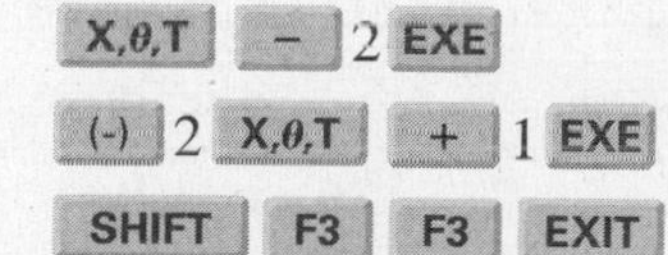

X,θ,T − 2 EXE

(-) 2 X,θ,T + 1 EXE

SHIFT F3 F3 EXIT

F6 F5 F5

Lesson 7.2

NAME ______________________________ DATE __________

Practice A

For use with pages 405–410

Solve for the indicated variable.

1. $5x + y = 8; y$

2. $2x - y = 4; y$

3. $x - 3y = 7; x$

4. $2x + 4y = 8; x$

5. $3x - 3y = -9; y$

6. $-\frac{1}{2}x + 5y = 3; x$

Tell which equation you would use to isolate a variable. Explain your reasoning.

7. $3x - y = 5$
$2x + y = 0$

8. $-2a + b = 7$
$3a + b = -8$

9. $2m + 5n = 14$
$2m - 3n = 6$

Use the substitution method to solve the linear system.

10. $y = x + 2$
$2x + y = 8$

11. $y = x - 1$
$2x - y = 0$

12. $2x + y = 3$
$y = 7$

13. $3x - y = -2$
$y = 2x + 3$

14. $x - 2y = 8$
$y = -4x + 5$

15. $y = -3x - 1$
$x - 3y = 3$

16. $x + y = -3$
$3x + y = 3$

17. $x - y = 4$
$x - 2y = 10$

18. $3x + y = 0$
$x - y = 4$

19. $3x - y = 9$
$2x + y = 6$

20. $x - 2y = 0$
$3x + y = 0$

21. $2x - y = 3$
$3x - y = 4$

22. ***Driving*** Your brother and sister took turns driving on a 580-mile trip that took 10 hours to complete. Your brother drove at a constant speed of 55 miles per hour and your sister drove at a constant speed of 60 miles per hour. Assign labels to the verbal model below. Write and solve an algebraic model. How long did each person drive?

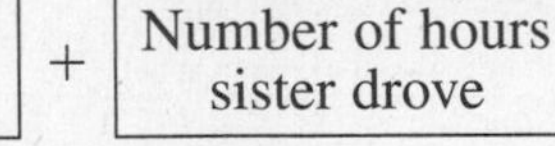

[Number of hours brother drove] + [Number of hours sister drove] = [Total number of hours]

[Brother's speed] · [Number of hours brother drove] + [Sister's speed] · [Number of hours sister drove] = [Total number of miles]

23. ***Dimensions of a Metal Sheet*** A rectangular hole 2 centimeters wide and x centimeters long is cut in a rectangular sheet of metal 5 centimeters wide and y centimeters long. The length of the hole is 8 centimeters less than the length of the metal sheet. After the hole is cut, the area of the remaining metal is 49 cm². Find the length of the hole and the length of the metal sheet.

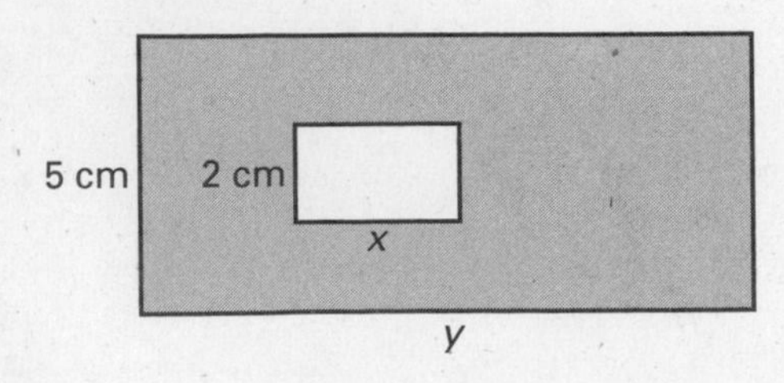

LESSON 7.2

NAME ____________________ DATE __________

Practice B

For use with pages 405–410

Solve for the indicated variable.

1. $5x + y = -8; y$
2. $6x - y = 4; y$
3. $x + 3y = 7; x$
4. $-2x + 4y = 8; x$
5. $-3x - 3y = 9; y$
6. $-\frac{1}{2}x + 5y = -3; x$

Tell which equation you would use to isolate a variable. Explain your reasoning.

7. $4x - y = -6$
   $2x + y = 0$
8. $2a + 4b = 10$
   $3a - b = 1$
9. $-m + 5n = 16$
   $-2m + 3n = 4$

Use the substitution method to solve the linear system.

10. $y = x + 3$
    $3x - y = 5$
11. $4x + y = 9$
    $y = -7$
12. $3x = 9$
    $-2x + y = -8$
13. $x - 2y = -13$
    $y = -2x - 6$
14. $x - y = 10$
    $5x - y = -6$
15. $4x + y = 2$
    $x - y = -17$
16. $-x + 3y = 4$
    $x + 6y = 14$
17. $3x + 2y = 8$
    $x + 4y = -4$
18. $x - 5y = -3$
    $4x - 3y = 5$
19. $2x + 5y = 4$
    $x + 5y = 7$
20. $\frac{1}{2}x + y = 2$
    $2x + 3y = 9$
21. $\frac{1}{3}x + \frac{5}{6}y = 1$
    $-\frac{1}{2}x - y = 1$

22. ***Mowing and Shoveling*** Last year you mowed grass and shoveled snow for 10 households. You earned \$200 per household mowing for the entire season and \$180 per household shoveling for the entire season. If you earned a total of \$1880 last year, how many households did you mow and shovel for? Assign labels to the verbal model below. Write and solve an algebraic model.

Number of households mow for	+	Number of households shovel for	=	Total number of households

Earnings per household mowing	·	Number of households mow for	+	Earnings per household shoveling	·	Number of households shovel for	=	Total earnings

23. ***Dimensions of a Metal Sheet*** A rectangular hole 2 centimeters wide and x centimeters long is cut in a rectangular sheet of metal $\frac{7}{2}$ centimeters wide and y centimeters long. The length of the hole is 1 centimeter less than the length of the metal sheet. After the hole is cut, the area of the remaining metal is 11 cm². Find the length of the hole and the length of the metal sheet.

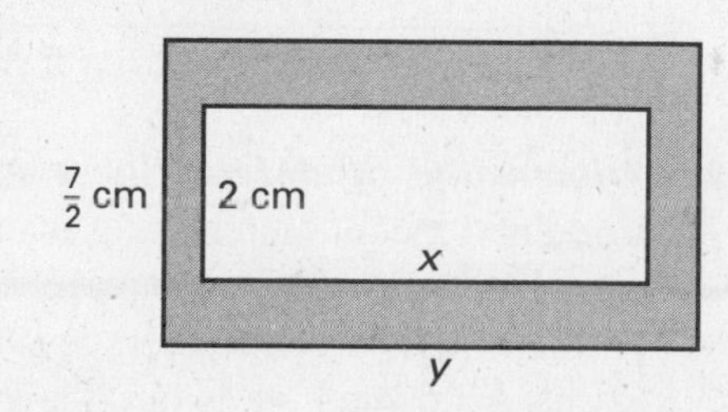

Lesson 7.2

LESSON 7.2

NAME ______________________ DATE __________

Practice C

For use with pages 405–410

Tell which equation you would use to isolate a variable. Explain your reasoning.

1. $3x - y = -15$
   $2x + y = 0$

2. $-2a + 4b = 6$
   $3a - b = 1$

3. $m - 5n = 18$
   $2m - 3n = -13$

Use the substitution method to solve the linear system.

4. $y = x - 4$
   $3x - y = -8$

5. $y = x + 3$
   $-7x - y = 1$

6. $4x + y = 12$
   $y = -8$

7. $-3x = 9$
   $2x + y = -13$

8. $x - 5y = -10$
   $y = -2x - 9$

9. $x - y = -7$
   $6x - y = -2$

10. $-3x + y = -18$
    $x - y = 14$

11. $-x + 5y = -7$
    $-x - 6y = 15$

12. $5x + 3y = 1$
    $x + 6y = 2$

13. $-3x + y = -3$
    $2x - 5y = -11$

14. $2x - 3y = -14$
    $3x - y = -7$

15. $\frac{1}{2}x + y = 3$
    $2x + 3y = 10$

16. $\frac{1}{3}x + \frac{2}{3}y = 12$
    $x + 9y = -6$

17. $x - \frac{3}{4}y = 1$
    $-2x + y = -\frac{3}{2}$

18. $\frac{1}{10}x - \frac{2}{5}y = 2$
    $\frac{1}{2}x - y = 10$

19. ***Mowing and Shoveling*** Last year you mowed grass and shoveled snow for 10 households. You earned \$200 per household mowing for the entire season and \$180 per household shoveling for the entire season. If you earned a total of \$1880 last year, how many households did you mow and shovel for?

20. ***Room Dimensions*** The area of the room shown below is 140 square feet. The perimeter of the room is 52 feet. Find x and y.

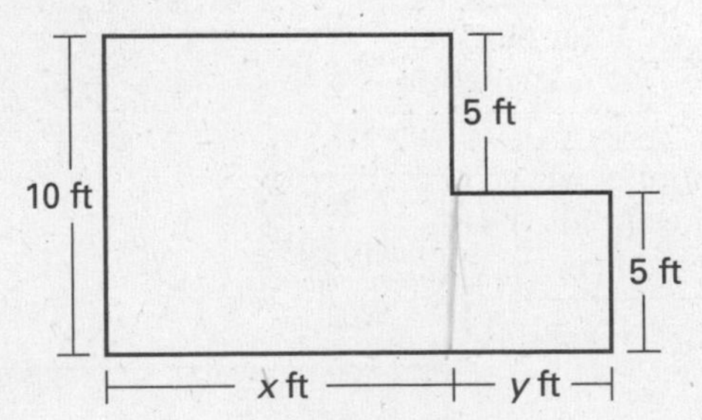

21. ***Dimensions of a Triangle*** The perimeter of an isosceles triangle is 16 inches. The area of the triangle is 12 square inches. What are the lengths of the sides of the isosceles triangle?

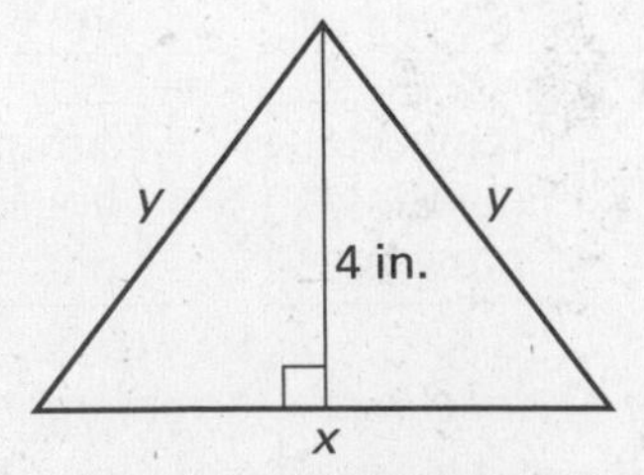

22. ***Dimensions of a Metal Sheet*** A rectangular hole $\frac{3}{4}$ centimeters wide and x centimeters long is cut in a rectangular sheet of metal $\frac{7}{2}$ centimeters wide and y centimeters long. The length of the hole is 5 centimeters less than the length of the metal sheet. After the hole is cut, the area of the remaining metal is 23 cm^2. Find the length of the hole and the length of the metal sheet.

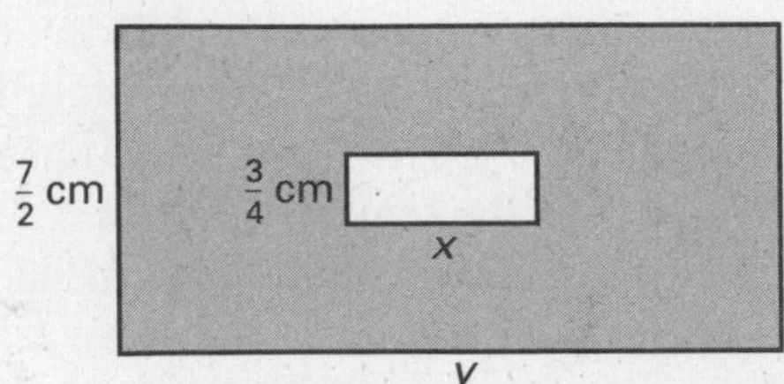

LESSON 7.2

NAME ______________________ DATE __________

Reteaching with Practice

For use with pages 405–410

GOAL **Use substitution to solve a linear system and model a real-life situation using a linear system**

EXAMPLE 1 *The Substitution Method*

Solve the linear system.

$x + y = 1$ Equation 1

$2x - 3y = 12$ Equation 2

SOLUTION

Solve for y in Equation 1.

$y = -x + 1$ Revised Equation 1

Substitute $-x + 1$ for y in Equation 2 and solve for x.

$2x - 3y = 12$ Write Equation 2.

$2x - 3(-x + 1) = 12$ Substitute $-x + 1$ for y.

$2x + 3x - 3 = 12$ Distribute the -3.

$5x - 3 = 12$ Simplify.

$5x = 15$ Add 3 to each side.

$x = 3$ Solve for x.

To find the value of y, substitute 3 for x in the revised Equation 1.

$y = -x + 1$ Write revised Equation 1.

$y = -3 + 1$ Substitute 3 for x.

$y = -2$ Solve for y.

The solution is $(3, -2)$.

Exercises for Example 1

Use the substitution method to solve the linear system.

1. $x + 2y = -5$
$4x - 3y = 2$

2. $3x - 2y = 4$
$x + 3y = 5$

3. $3x + y = -2$
$x + 3y = 2$

Lesson 7.2

NAME ______________________ DATE ________

Reteaching with Practice

For use with pages 405–410

EXAMPLE 2 Writing and Using a Linear System

An investor bought 225 shares of stock, stock A at $50 per share and stock B at $75 per share. If $13,750 worth of stock was purchased, how many shares of each kind did the investor buy?

SOLUTION

Verbal Model

[Amount of stock A] + [Amount of stock B] = [Total amount of stock]

[Price of stock A] · [Amount of stock A] + [Price of stock B] · [Amount of stock B] = [Total investment]

Labels

Amount of stock A $= x$		(shares)
Amount of stock B $= y$		(shares)
Total amount of stock $= 225$		(shares)
Price of stock A $= 50$		(dollars per share)
Price of stock B $= 75$		(dollars per share)
Total investment $= 13{,}750$		(dollars)

Algebraic Model

$x + y = 225$ — Equation 1 (shares)

$50x + 75y = 13{,}750$ — Equation 2 (dollars)

Solve for y in Equation 1.

$y = -x + 225$ — Revised Equation 1

Substitute $-x + 225$ for y in Equation 2 and solve for x.

$50x + 75y = 13{,}750$ — Write Equation 2.

$50x + 75(-x + 225) = 13{,}750$ — Substitute $-x + 225$ for y.

$50x - 75x + 16{,}875 = 13{,}750$ — Distribute the 75.

$-25x = -3125$ — Simplify.

$x = 125$ — Solve for x.

To find the value of y, substitute 125 for x in the revised Equation 1.

$y = -x + 225$ — Write revised Equation 1.

$y = -125 + 225$ — Substitute 125 for x.

$y = 100$ — Solve for y.

The solution is (125, 100).

Exercises for Example 2

4. Rework Example 2 if the investor bought 200 shares of stock.
5. Rework Example 2 if $16,250 worth of stock was purchased.

LESSON 7.2

NAME ______________________________ DATE __________

Quick Catch-Up for Absent Students

For use with pages 405–410

The items checked below were covered in class on (date missed) __________

Lesson 7.2: Solving Linear Systems by Substitution

____ **Goal 1:** Use substitution to solve a linear system. (pp. 405–406)

Material Covered:

____ Student Help: Study Tip

____ Example 1: The Substitution Method

____ Student Help: Study Tip

____ Example 2: The Substitution Method

____ **Goal 2:** Model a real-life situation using a linear system. (p. 407)

Material Covered:

____ Example 3: Writing and Using a Linear System

____ Other (specify) ______________________________

____

Homework and Additional Learning Support

____ Textbook (specify) pp. 408–410

____

____ *Reteaching with Practice* worksheet (specify exercises) __________

____ *Personal Student Tutor* for Lesson 7.2

Lesson 7.2

NAME ______________________ DATE __________

Interdisciplinary Application

For use with pages 405–410

Amphibians

BIOLOGY An amphibian is an animal with scale-less skin that–with a few exceptions–lives part of its life in water and part on land. There are about 4000 kinds of amphibians, and they make up one of the classes of vertebrates. Zoologists divide amphibians into three groups: (1) frogs and toads; (2) salamanders; and (3) caecilians.

Amphibians are cold-blooded–that is, their body temperature stays about the same as the temperature of their surroundings. Those that live in regions with harsh winters hibernate during the cold weather. Many of those that live in warm, dry climates estivate–that is, become inactive during summer.

Amphibians live on every continent except Antarctica. They generally live in moist habitats near ponds, lakes, or streams. Most amphibians eat insects. In some areas of the world, amphibians are quite numerous, and they play an important role in maintaining the balance of nature. Amphibians aid people by eating insects and insect larvae that destroy crops and carry disease.

In Exercises 1-4, use the following information.

There are many amphibians in your biology classroom. Your teacher tells you there are two kinds of amphibians, frogs and salamanders, 44 in all. The chart on the wall says to feed each frog 5 crickets a day and each salamander 3 crickets a day. You know that 182 crickets are used in one day.

1. Write a linear system to represent the situation. One equation should show the total number of amphibians. The other should show the total number of crickets used for feeding in one day.
2. Using substitution, find the number of frogs in the classroom.
3. Using your answer from Exercise 2, find the number of salamanders in the classroom.
4. Check your results by graphing your linear system.

LESSON 7.2

NAME ______________________ DATE __________

Challenge: Skills and Applications

For use with pages 405–410

In Exercises 1–2, use the substitution method to solve the linear system.

1. $3x + 2y = 5$
$5x + 3y = 7$

2. $4x - 5y = 22$
$2x - 7y = 20$

In Exercises 3–4, use the method shown in the following example to solve the linear system.

Example: $2x - y^2 = -11$
$x + 3y^2 = 26$

Solution: Let $u = y^2$. Then the system becomes

$2x - u = -11$

$x + 3u = 26$

Solving this system by substitution, you get $x = -1$ and $u = 9$.
If $y^2 = 9$, $y = 3$, or $y = -3$, because $3^2 = 9$ and $(-3)^2 = 9$.
Thus, the solution is $(-1, 3)$ and $(-1, -3)$.

3. $3x^2 + y = 21$
$x^2 + 2y = 22$

4. $4\left(\frac{1}{x}\right) + \frac{1}{y} = 11$
$2\left(\frac{1}{x}\right) - 5\left(\frac{1}{y}\right) = 11$

In Exercises 5–7, use the following information.

Galvanized nails are sold by the pound. Robert Long bought \$2.00 worth of six-penny nails and \$3.00 worth of eight-penny nails for a total weight of 8 pounds. Later he bought \$1.00 worth of six-penny nails and \$6.00 worth of eight-penny nails for a total weight of 10 pounds. Robert wants to know how much per pound each type of nail costs.

5. Write a linear system to represent Robert's purchases, where x represents the cost per pound of six-penny nails and y represents the cost per pound of eight-penny nails.

6. Let $u = \frac{1}{x}$ and $v = \frac{1}{y}$. Substitute into the system from Exercise 5 and solve for u and v.

7. How much per pound does each type of nail cost?

Lesson 7.2

LESSON

TEACHER'S NAME ______________________ CLASS ______ ROOM ______ DATE ______

Lesson Plan

2-day lesson (See *Pacing the Chapter,* TE pages 394C–394D) For use with pages 411–417

GOALS
1. Use linear combinations to solve a system of linear equations.
2. Model a real-life problem using a system of linear equations.

State/Local Objectives ______________________________

____

✓ **Check the items you wish to use for this lesson.**

STARTING OPTIONS

____ Homework Check: TE page 408; Answer Transparencies
____ Warm-Up or Daily Homework Quiz: TE pages 411 and 410, CRB page 38, or Transparencies

TEACHING OPTIONS

____ Motivating the Lesson: TE page 412
____ Lesson Opener (Application): CRB page 39 or Transparencies
____ Examples: Day 1: 1–3, SE pages 411–412; Day 2: 4, SE page 413
____ Extra Examples: Day 1: TE page 412 or Transp.; Day 2: TE page 413 or Transp.; Internet
____ Closure Question: TE page 413
____ Guided Practice: SE page 414; Day 1: Exs. 1–8; Day 2: none

APPLY/HOMEWORK

Homework Assignment

____ Basic Day 1: 8–30 even; Day 2: 32–42 even, 43, 44, 49–51, 53–55, 60, 65, 70; Quiz 1: 1–10
____ Average Day 1: 8–30 even; Day 2: 32–42 even, 43–51, 53–55, 60, 65, 70; Quiz 1: 1–10
____ Advanced Day 1: 8–30 even; Day 2: 32–42 even, 43–56, 60, 65, 70; Quiz 1: 1–10

Reteaching the Lesson

____ Practice Masters: CRB pages 40–42 (Level A, Level B, Level C)
____ Reteaching with Practice: CRB pages 43–44 or Practice Workbook with Examples
____ Personal Student Tutor

Extending the Lesson

____ Applications (Real-Life): CRB page 46
____ Math & History: SE page 417; CRB page 47; Internet
____ Challenge: SE page 416; CRB page 48 or Internet

ASSESSMENT OPTIONS

____ Checkpoint Exercises: Day 1: TE page 412 or Transp.; Day 2: TE page 413 or Transp.
____ Daily Homework Quiz (7.3): TE page 416, CRB page 52, or Transparencies
____ Standardized Test Practice: SE page 416; TE page 416; STP Workbook; Transparencies
____ Quizzes (7.1–7.3): SE page 417; CRB page 49

Notes ______________________________

____

____

LESSON 7.3

TEACHER'S NAME ____________ CLASS ______ ROOM ______ DATE ______

Lesson Plan for Block Scheduling

1-day lesson (See *Pacing the Chapter,* TE pages 394C–394D) For use with pages 411–417

GOALS
1. Use linear combinations to solve a system of linear equations.
2. Model a real-life problem using a system of linear equations.

State/Local Objectives ____________

CHAPTER PACING GUIDE	
Day	Lesson
1	7.1 (all); 7.2 (all)
2	**7.3 (all)**
3	7.4 (all)
4	7.5 (all); 7.6 (begin)
5	7.6 (end); Review Ch. 7
6	Assess Ch. 7; 8.1 (begin)

✓ **Check the items you wish to use for this lesson.**

STARTING OPTIONS

____ Homework Check: TE page 408; Answer Transparencies
____ Warm-Up or Daily Homework Quiz: TE pages 411 and 410, CRB page 38, or Transparencies

TEACHING OPTIONS

____ Motivating the Lesson: TE page 412
____ Lesson Opener (Application): CRB page 39 or Transparencies
____ Examples 1–4: SE pages 411–413
____ Extra Examples: TE pages 412–413 or Transparencies; Internet
____ Closure Question: TE page 413
____ Guided Practice Exercises: SE page 414

APPLY/HOMEWORK

Homework Assignment

____ Block Schedule: 8–42 even, 43–51, 53–55, 60, 65, 70; Quiz 1: 1–10

Reteaching the Lesson

____ Practice Masters: CRB pages 40–42 (Level A, Level B, Level C)
____ Reteaching with Practice: CRB pages 43–44 or Practice Workbook with Examples
____ Personal Student Tutor

Extending the Lesson

____ Applications (Real-Life): CRB page 46
____ Math & History: SE page 417; CRB page 47; Internet
____ Challenge: SE page 416; CRB page 48 or Internet

ASSESSMENT OPTIONS

____ Checkpoint Exercises: TE pages 412–413 or Transparencies
____ Daily Homework Quiz (7.3): TE page 416, CRB page 52, or Transparencies
____ Standardized Test Practice: SE page 416; TE page 416; STP Workbook; Transparencies
____ Quizzes (7.1–7.3): SE page 417; CRB page 49

Notes ____________

NAME ____________________ DATE ________

Available as a transparency

WARM-UP EXERCISES

For use before Lesson 7.3, pages 411–417

Give the opposite of each number.

1. 3

2. 1

3. $-\frac{1}{2}$

Simplify.

4. $3(2x - 7y)$

5. $-2(5x + y)$

DAILY HOMEWORK QUIZ

For use after Lesson 7.2, pages 405–410

Solve the linear system.

1. $-3x + 2y = -11$
$5x - y = 23$

2. $a + 4b = 8$
$2a + 3b = 1$

3. $5p - 4q = -3$
$2p - q = -3$

4. $m = 7n$
$3m + 8n = -29$

5. A sightseeing boat charges $5 for children and $8 for adults. On its first trip of the day, it collected $439 for 71 paying passengers. How many children and how many adults were there?

NAME ______________________ DATE __________

Available as a transparency

Application Lesson Opener

For use with pages 411–417

Randy has some \$1 bills and \$5 bills in his wallet. He has 15 bills in all. He counts the money and finds he has \$47. How many of each type of bill does Randy have?

Let x represent the number of \$1 bills and y represent the number of \$5 bills. The system of equations below is an algebraic model for this problem.

$$x + y = 15$$
$$x + 5y = 47$$

1. Add the equations by adding the like terms. Describe the result. Can you solve the resulting equation? Why or why not?

2. Subtract the equations by subtracting the like terms. Describe the result. Can you solve the resulting equation? Why or why not?

Beka is supposed to work the same number of hours each week. One week, she worked more hours than usual. The next week, she took the same number of hours off. If she worked 42 hours the first week and 34 hours the second week, how many hours is Beka supposed to work?

Let x represent the number of hours Beka is supposed to work and y represent the change from her usual number of hours. The system of equations is an algebraic model for this problem.

$$x + y = 42$$
$$x - y = 34$$

3. Add the equations by adding the like terms. Describe the result. Can you solve the resulting equation? Why or why not?

4. Subtract the equations by subtracting the like terms. Describe the result. Can you solve the resulting equation? Why or why not?

NAME ______________________________ DATE __________

Practice A

For use with pages 411–417

Use linear combinations to solve the system of linear equations. Use the graph to check your solution.

1. $2x + y = 4$
$x - y = 2$

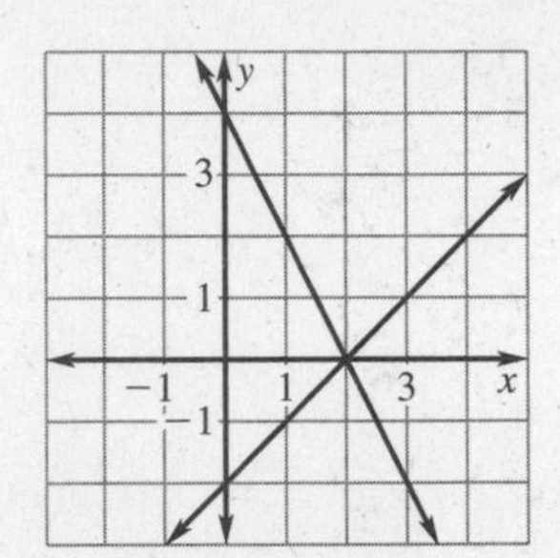

2. $x + 3y = 2$
$-x + 2y = 3$

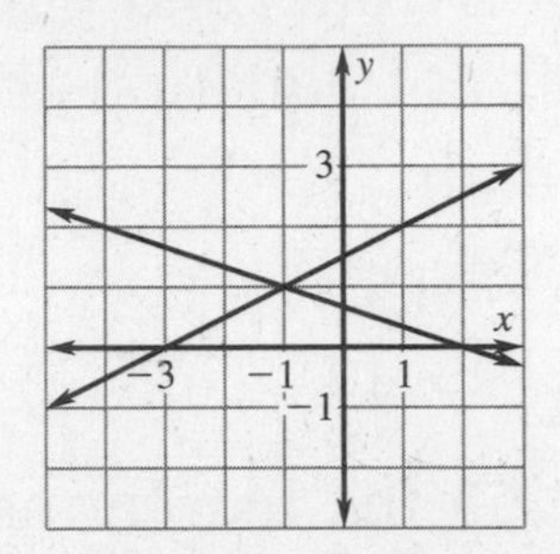

3. $2x - y = 2$
$4x + 3y = 24$

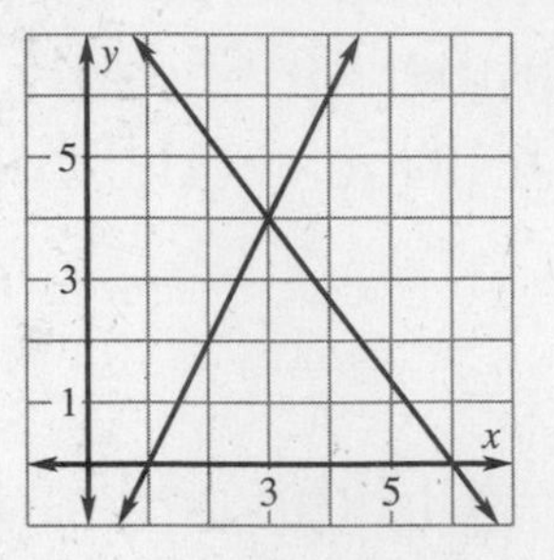

Use linear combinations to solve the system of linear equations.

4. $x + y = 5$
$x - y = 7$

5. $x - 2y = 8$
$-x + 3y = -5$

6. $x - 4y = 14$
$-x + 3y = -11$

7. $2x - y = -3$
$-5x + y = 9$

8. $3x + y = 6$
$-3x + 4y = 9$

9. $2x - 3y = -16$
$x + 3y = 10$

10. $x + 3y = -3$
$x - 4y = 11$

11. $-2x + 3y = 14$
$x - 4y = -12$

12. $5x + 2y = 5$
$3x + y = 2$

13. $2x - y = 1$
$2x + 5y = -5$

14. $4x - 5y = -18$
$5x + 4y = -2$

15. $2x + 5y = -22$
$4x - 3y = 8$

16. $4x = -3 + y$
$y = -6x - 7$

17. $x = 2y + 9$
$2y = 3x - 19$

18. $5y - 3x = -4$
$3x + 4y = 13$

19. $4x = 5y + 6$
$3y + 2x = -8$

20. $3y = 5x + 15$
$6x = 2y - 18$

21. $\frac{1}{2}x = 4y$
$5y - x = -3$

Electricians **In Exercises 22–24, use the following information.**

The yellow pages identify two different local electrical businesses. Business A charges \$50 for a service call, plus an additional \$40 per hour for labor. Business B charges \$30 for a service call, plus an additional \$45 per hour for labor.

22. Let x represent the number of hours of labor and let y represent the total charge. Write a system of equations you could solve to find the lengths of a service call for which both businesses charge the same amount.

23. Solve the system.

24. Which company would you use? Why?

Travel Agency **In Exercises 25 and 26, use the following information.**

A travel agency offers two Boston outings. Plan A includes hotel accommodations for three nights and two pairs of baseball tickets worth \$645. Plan B includes hotel accommodations for five nights and four pairs of baseball tickets worth \$1135.

25. Let x represent the cost of one night's hotel accommodation and let y represent the cost of one pair of baseball tickets. Write a system of equations you could solve to find the cost of one night's hotel accommodation and one pair of baseball tickets.

26. Solve the system.

LESSON 7.3

NAME ______________________________ DATE __________

Practice B

For use with pages 411–417

Use linear combinations to solve the system of linear equations.

1. $x + y = 11$
   $x - y = 7$

2. $x - 2y = 8$
   $-x + 3y = -15$

3. $3x + y = -8$
   $-3x + 4y = -2$

4. $2x - 4y = 14$
   $-2x + 3y = -11$

5. $\frac{1}{2}x - y = -3$
   $-5x + y = 12$

6. $7.5x - 1.2y = -2.7$
   $-1.5x + 1.2y = -3.3$

7. $x + 2y = -3$
   $x - 4y = 15$

8. $-x - 5y = 30$
   $2x - 7y = 25$

9. $-x + 8y = 16$
   $3x + 4y = 36$

10. $4x - 3y = -3$
    $4x + 5y = 5$

11. $4x + 5y = -2$
    $5x - 4y = -23$

12. $9x - 4y = -18$
    $-3x + 8y = 6$

13. $4x = -11 + y$
    $y = -6x - 9$

14. $x = 2y - 3$
    $2y = 3x + 13$

15. $4y = 15 - 3x$
    $2y = 3x + 21$

16. $4x = 5y - 14$
    $3y - 8x = -14$

17. $5x = 4y - 30$
    $2x + 3y = -12$

18. $\frac{2}{3}y = 10 + 4x$
    $5x = \frac{1}{3}y - 8$

Electricians **In Exercises 19-21, use the following information.**

The yellow pages identify two different local electrical businesses. Business A charges $50 for a service call, plus an additional $36 per hour for labor. Business B charges $35 for a service call, plus an additional $39 per hour for labor.

19. Let x represent the number of hours of labor and let y represent the total charge. Write a system of equations you could solve to find the length of a service call for which both businesses charge the same amount.

20. Solve the system.

21. Which company would you use? Why?

Travel Agency **In Exercises 22 and 23, use the following information.**

A travel agency offers two Boston outings. Plan A includes hotel accommodations for three nights and two pairs of baseball tickets worth $518. Plan B includes hotel accommodations for five nights and four pairs of baseball tickets worth $907.

22. Let x represent the cost of one night's hotel accommodation and let y represent the cost of one pair of baseball tickets. Write a system of equations you could solve to find the cost of one night's hotel accommodation and one pair of baseball tickets.

23. Solve the system.

Highway Project **In Exercises 24 and 25, use the following information.**

There are sixteen workers employed on a highway project, some at $200 per day and some at $165 per day. The daily payroll is $2745.

24. Let x represent the number of $200 per day workers and let y represent the number of $165 per day workers. Write a system of equations to find the number of workers employed at each wage.

25. Solve the system.

NAME ______________________________ DATE __________

Practice C

For use with pages 411–417

Use linear combinations to solve the system of linear equations.

1. $x + y = 15$
$x - y = 17$

2. $x - 2y = 8$
$-x + 5y = -5$

3. $3x + y = 16$
$-3x + 4y = 19$

4. $4x - 4y = 14$
$-4x + 3y = -11$

5. $\frac{2}{3}x - y = -4$
$-5x + y = -9$

6. $-5.4x - 0.8y = -30.2$
$4.2x + 0.8y = 24.2$

7. $x + 3y = -7$
$x - 5y = 9$

8. $-x - 5y = -20$
$2x - 7y = -45$

9. $2x - y = 16$
$3x + 5y = 11$

10. $5x + 4y = -10$
$3x + 4y = -6$

11. $6x - 7y = 9$
$7x + 6y = -32$

12. $-9x + 4y = -12$
$3x - 8y = 24$

13. $4x = -11 + y$
$y = -7x - 11$

14. $x = 4y + 13$
$4y = 3x - 23$

15. $8y = 7x + 32$
$6y = 24 - 5x$

16. $2y = 1 - 3x$
$3y = -(1 + 4x)$

17. $\frac{1}{2}x - \frac{1}{5}y = 1$
$y - \frac{1}{3}x = 8$

18. $0.2x = 1.1 - 0.3y$
$0.3y = 1.8 - 0.3x$

Electricians **In Exercises 19-21, use the following information.**

The yellow pages identify two different local electrical businesses. Business A charges \$50 for a service call, plus an additional \$39 per hour for labor. Business B charges \$32 for a service call, plus an additional \$45 per hour for labor.

19. Let x represent the number of hours of labor and let y represent the total charge. Write a system of equations you could solve to find the length of a service call for which both businesses charge the same amount.

20. Solve the system.

21. Which company would you use? Why?

Travel Agency **In Exercises 22 and 23, use the following information.**

A travel agency offers two Boston outings. Plan A includes hotel accommodations for three nights and two pairs of baseball tickets worth \$556.40. Plan B includes hotel accommodations for five nights and four pairs of baseball tickets worth \$973.

22. Let x represent the cost of one night's hotel accommodation and let y represent the cost of one pair of baseball tickets. Write a system of equations you could solve to find the cost of one night's hotel accommodation and one pair of baseball tickets.

23. Solve the system.

24. ***Highway Project*** There are sixteen workers employed on a highway project, some at \$200 per day and some at \$165 per day. The daily payroll is \$2745. Find the number of workers employed at each wage.

25. ***Concert Tickets*** In one day, the Ticket-Taker Agency sold 395 tickets for a concert, some at \$28 per ticket and some at \$22 per ticket. If the agency collected \$10,130 that day, find the number of tickets sold at each price.

NAME ______________________________ DATE ____________

Reteaching with Practice

For use with pages 411–417

GOAL **Use linear combinations to solve a system of linear equations and model a real-life problem using a system of linear equations**

VOCABULARY

A **linear combination** of two equations is an equation obtained by adding one of the equations (or a multiple of one of the equations) to the other equation.

EXAMPLE 1 *Using Multiplication First*

Solve the linear system. $4x - 3y = 11$ Equation 1

$3x + 2y = -13$ Equation 2

SOLUTION

The equations are arranged with like terms in columns. You can get the coefficients of y to be opposites by multiplying the first equation by 2 and the second equation by 3.

$4x - 3y = 11$ Multiply by 2. $8x - 6y = 22$

$3x + 2y = -13$ Multiply by 3. $\underline{9x + 6y = -39}$

$17x = -17$ Add the equations.

$x = -1$ Solve for x.

Substitute -1 for x in the second equation and solve for y.

$3x + 2y = -13$ Write Equation 2.

$3(-1) + 2y = -13$ Substitute -1 for x.

$-3 + 2y = -13$ Simplify.

$y = -5$ Solve for y.

The solution is $(-1, -5)$.

Exercises for Example 1

Use linear combinations to solve the system of linear equations.

1. $x + 2y = 5$
$3x - 2y = 7$

2. $x + y = 1$
$2x - 3y = 12$

3. $x - y = -4$
$x + 2y = 5$

NAME ____________________ DATE ________

Reteaching with Practice

For use with pages 411–417

EXAMPLE 2 Writing and Using a Linear System

A pharmacy mailed 300 advertisements, smaller ads requiring $.33 postage and larger ads requiring $.55 postage. If the total cost of postage was $121, find the number of advertisements mailed at each rate.

SOLUTION

Verbal Model

[Number of smaller ads] + [Number of larger ads] = [Total number of ads]

[Postage for smaller ads] · [Number of smaller ads] + [Postage for larger ads] · [Number of larger ads] = [Total cost of postage]

Labels

Number of smaller ads = x (ads)

Number of larger ads = y (ads)

Total number of ads = 300 (ads)

Postage for smaller ads = 0.33 (dollars per ad)

Postage for larger ads = 0.55 (dollars per ad)

Total cost of postage = 121 (dollars)

Algebraic Model

$x + y = 300$ Equation 1 (ads)

$0.33x + 0.55y = 121$ Equation 2 (dollars)

Use linear combinations to solve for y.

$-0.33x - 0.33y = -99$ Multiply Equation 1 by -0.33.

$0.33x + 0.55y = 121$ Write Equation 2.

$0.22y = 22$ Add the equations.

$y = 100$ Solve for y.

Substitute 100 for y in Equation 1 and solve for x.

$x + y = 300$ Write Equation 1.

$x + 100 = 300$ Substitute 100 for y.

$x = 200$ Solve for x.

The solution is (200, 100). The pharmacy mailed 200 smaller ads and 100 larger ads.

Exercises for Example 2

4. Rework Example 2 if the total cost of postage was $154.

5. Rework Example 2 if the pharmacy mailed 320 advertisements.

NAME ______________________ DATE __________

Quick Catch-Up for Absent Students

For use with pages 411–417

The items checked below were covered in class on (date missed) __________

Lesson 7.3: Solving Linear Systems by Linear Combinations

____ **Goal 1:** Use linear combinations to solve a system of linear equations. (pp. 411–412)

Material Covered:

____ Example 1: Using Addition

____ Student Help: Study Tip

____ Example 2: Using Multiplication First

____ Example 3: Arranging Like Terms in Columns

Vocabulary:

linear combination, p. 411

____ **Goal 2:** Model a real-life problem using a system of linear equations. (p. 413)

Material Covered:

____ Example 4: Writing and Using a Linear System

____ Other (specify) ______________________

____

Homework and Additional Learning Support

____ Textbook (specify) pp. 414–417

____

____ Internet: Extra Examples at www.mcdougallittell.com

____ *Reteaching with Practice* worksheet (specify exercises) ______________

____ *Personal Student Tutor* for Lesson 7.3

NAME ______________________ DATE ________

Real-Life Application: When Will I Ever Use This?

For use with pages 411–417

The Juan Fernandez Islands

Juan Fernandez is the name of a group of three islands that lie off the West Coast of Chile in the Pacific Ocean. Juan Fernandez, a Spanish explorer, discovered the islands in 1574. These islands, Robinson Crusoe, Santa Clara, and Alejandro Selkirk, are part of Chile. The island group has an area of 56 square miles. Most of the people fish for a living. The islands' waters are well known for the lobsters caught there.

Only Robinson Crusoe, the second largest island of the group, has a permanent population. About 500 Spanish-speaking people live there. It is famous because this is the island where Alexander Selkirk stayed alone for more than four years (1704-1709). The English writer Daniel Defoe partly based his Robinson Crusoe on Selkirk's adventures.

In Exercises 1-4, use the following information.

A ship carrying people to the island of Robinson Crusoe leaves the port city of La Serena, Chile. The ship travels on a straight line represented by the equation $10y - 7x = 127$. Another ship leaves the port city of Valdivia with supplies for the island dwellers. The equation $1.1x + y = -7.1$ represents the path taken by the ship leaving from Valdivia. Ships navigate with the use of a coordinate system because there is nothing to look at in the middle of an ocean. Will the ships end up at the same coordinates, the island of Robinson Crusoe?

1. Write the equations in standard form.
2. Solve the linear system using linear combinations.
3. Interpret your answer? (What do the values of x and y tell you?)
4. Graph the linear system to check your answer to Exercise 2.

NAME ______________________________ DATE __________

Math and History Application

For use with page 417

HISTORY *Chu Chang Suan Shu (Nine Chapters on the Mathematical Art)* was written about 2000 years ago during the Han Dynasty. The Chinese mathematician Shang Tshang probably assembled this text by making use of older works already in existence. *Chu Chang Suan Shu* is considered one of the oldest mathematical textbooks. The book is a collection of many mathematicians' efforts thoughout the centuries. The original book is believed to have been destroyed in the Burning of the Books of 213 B.C. *Nine Chapters on the Mathematical Art* is also the first textbook to show methods for solving simultaneous linear equations.

The text contains 246 questions with general rules for solutions. The topics focus upon applied mathematics in engineering and administration. Some of the chapter titles are surveying land, engineering works, and impartial taxation. The problems are varied and include weights and measure, payment for livestock, and construction of canals.

MATH Chapter eight is titled *Fang cheng* (rectangular arrays). The topics in this chapter include simultaneous linear equations, the concept of positive and negative numbers, and addition and subtraction of positive and negative numbers. The setup and solutions for the simultaneous linear equations are similar those in your textbook.

1. Solve the system of equations.

   $7x - 8y = -21$

   $3x + 2y = 29$

2. You plant a 4-foot maple tree that grows at a rate of 6 inches per year and a 2-foot birch tree that grows at a rate of 8 inches per year. In how many years after planting will the two trees be the same height? How tall will each tree be?

3. Three pieces of a metal A and two pieces of a metal B combined weigh 100 pounds. The difference between one piece of metal A and one piece of metal B is five pounds. How much does one piece of metal A weigh? How much does one piece of metal B weigh?

NAME ______________________ DATE __________

Challenge: Skills and Applications

For use with pages 411–417

In Exercises 1–2, use linear combinations to solve the system of equations.

Example: $-5\left(\frac{1}{x}\right) + 2\left(\frac{1}{y}\right) = -3$

$5\left(\frac{1}{x}\right) + 3\left(\frac{1}{y}\right) = 33$

Solution: Let $u = \frac{1}{x}$ and $v = \frac{1}{y}$. Substituting, the system becomes

$-5u + 2v = -3$

$5u + 3v = 33$

By adding, you get $v = 6$, $u = 3$. Therefore $3 = \frac{1}{x}$ and $6 = \frac{1}{y}$ or $x = \frac{1}{3}$ and $y = \frac{1}{6}$.

The solution is $\left(\frac{1}{3}, \frac{1}{6}\right)$.

1. $4\left(1 - \frac{1}{x}\right) - 3y = 11$

$10\left(1 - \frac{1}{x}\right) - 3y = 14$

2. $3x = y^3 - 6$

$8x = y^3 + 29$

In Exercises 3–8, use linear combinations to solve the system of equations for *x* and *y* in terms of the other variables.

3. $5ax + 3y = 5$

$2ax + 3y = 11$

4. $3x + 6y = 9a$

$5x - 6y = 7a$

5. $4x + ay = 7$

$-2x + ay = -11$

6. $x + by = 8$

$4x + 3by = 22$

7. $-ax + 2y = 5$

$4ax - 11y = -8$

8. $ax - 3by = 12$

$2ax - by = 19$

LESSON 7.3

NAME ______________________ DATE ________

Quiz 1

For use after Lessons 7.1–7.3

1. Decide whether $(-4, 2)$ is a solution of the system. *(Lesson 7.1)*

$3x + 2y = -8$

$2x - y = -6$

2. Graph and check to solve the linear system. *(Lesson 7.1)*

$5x - 2y = 4$

$-x + 4y = -8$

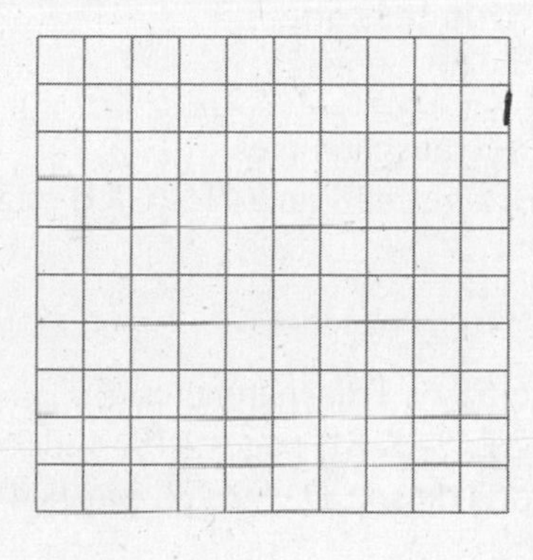

3. Use the substitution method to solve the linear system. *(Lesson 7.2)*

$2x - 2y = 12$

$x - 5y = 2$

4. Use linear combinations to solve the linear sytem. *(Lesson 7.3)*

$x = -\frac{1}{4}y + 2$

$5y - 4x = 16$

5. You have \$160 and save \$7 per week. Your friend has \$210 and saves \$5 per week. After how many weeks will each of you have saved the same amount of money? *(Lessons 7.1–7.3)*

Answers

1. ____________

2. ____________

Use grid at left.

3. ____________

4. ____________

5. ____________

Lesson 7.3

LESSON
7.4

TEACHER'S NAME ______________________ CLASS ______ ROOM ______ DATE ______

Lesson Plan

2-day lesson (See *Pacing the Chapter,* TE pages 394C–394D) For use with pages 418–424

1. Choose the best method to solve a system of linear equations.
2. Use a system to model real-life problems.

State/Local Objectives ____

____

✓ **Check the items you wish to use for this lesson.**

STARTING OPTIONS

____ Homework Check: TE page 414; Answer Transparencies
____ Warm-Up or Daily Homework Quiz: TE pages 418 and 416, CRB page 52, or Transparencies

TEACHING OPTIONS

____ Motivating the Lesson: TE page 419
____ Lesson Opener (Application): CRB page 53 or Transparencies
____ Examples: Day 1: 1, SE page 418; Day 2: 2–3, SE pages 419–420
____ Extra Examples: Day 1: TE page 419 or Transp.; Day 2: TE pages 419–420 or Transp.
____ Closure Question: TE page 420
____ Guided Practice: SE page 421; Day 1: Exs. 1–5; Day 2: Exs. 6–9

APPLY/HOMEWORK

Homework Assignment

____ Basic Day 1: 10–38 even; Day 2: 40–48 even, 49, 50, 61–66, 68–75
____ Average Day 1: 10–38 even; Day 2: 40–48 even, 49–51, 61–66, 68–75
____ Advanced Day 1: 10–38 even; Day 2: 40–48 even, 49–51, 61–75

Reteaching the Lesson

____ Practice Masters: CRB pages 54–56 (Level A, Level B, Level C)
____ Reteaching with Practice: CRB pages 57–58 or Practice Workbook with Examples
____ Personal Student Tutor

Extending the Lesson

____ Cooperative Learning Activity: CRB page 60
____ Applications (Interdisciplinary): CRB page 61
____ Challenge: SE page 424; CRB page 62 or Internet

ASSESSMENT OPTIONS

____ Checkpoint Exercises: Day 1: TE page 419 or Transp.; Day 2: TE pages 419–420 or Transp.
____ Daily Homework Quiz (7.4): TE page 424, CRB page 65, or Transparencies
____ Standardized Test Practice: SE page 424; TE page 424; STP Workbook; Transparencies

Notes ____

____

____

LESSON 7.4

TEACHER'S NAME ______________ CLASS ______ ROOM ______ DATE ______

Lesson Plan for Block Scheduling

1-day lesson (See *Pacing the Chapter,* TE pages 394C–394D)

For use with pages 418–424

GOALS
1. **Choose the best method to solve a system of linear equations.**
2. **Use a system to model real-life problems.**

State/Local Objectives ______________

CHAPTER PACING GUIDE	
Day	**Lesson**
1	7.1 (all); 7.2 (all)
2	7.3 (all)
3	**7.4 (all)**
4	7.5 (all); 7.6 (begin)
5	7.6 (end); Review Ch. 7
6	Assess Ch. 7; 8.1 (begin)

✓ **Check the items you wish to use for this lesson.**

STARTING OPTIONS

____ Homework Check: TE page 414; Answer Transparencies
____ Warm-Up or Daily Homework Quiz: TE pages 418 and 416, CRB page 52, or Transparencies

TEACHING OPTIONS

____ Motivating the Lesson: TE page 419
____ Lesson Opener (Application): CRB page 53 or Transparencies
____ Examples 1–3: SE pages 418–420
____ Extra Examples: TE pages 419–420 or Transparencies
____ Closure Question: TE page 420
____ Guided Practice Exercises: SE page 421

APPLY/HOMEWORK

Homework Assignment

____ Block Schedule: 10–48 even, 49–51, 61–66, 68–76

Reteaching the Lesson

____ Practice Masters: CRB pages 54–56 (Level A, Level B, Level C)
____ Reteaching with Practice: CRB pages 57–58 or Practice Workbook with Examples
____ Personal Student Tutor

Extending the Lesson

____ Cooperative Learning Activity: CRB page 60
____ Applications (Interdisciplinary): CRB page 61
____ Challenge: SE page 424; CRB page 62 or Internet

ASSESSMENT OPTIONS

____ Checkpoint Exercises: TE pages 419–420 or Transparencies
____ Daily Homework Quiz (7.4): TE page 424, CRB page 65, or Transparencies
____ Standardized Test Practice: SE page 424; TE page 424; STP Workbook; Transparencies

Notes ______________

LESSON 7.4

NAME ______________________ DATE ________

Available as a transparency

Warm-Up Exercises

For use before Lesson 7.4, pages 418–424

1. Solve the linear system using substitution.

$2x - y = 7$

$3x + 3y = -3$

2. Solve the linear system using the graph-and-check method.

$y = x$

$y = -2x + 3$

3. Solve the linear system using linear combinations.

$-3x + 4y = -4$

$3x - 6y = 6$

Daily Homework Quiz

For use after Lesson 7.3, pages 411–417

Solve the system.

1. $-3x + 5y = -7$

$3x - 8y = 1$

2. $7m + n = -2$

$3m + n = 2$

3. $8e - 3f = 12$

$2e + 7f = -28$

4. $10g + 3h = 10$

$-12g - 6h = -24$

5. $5p = 1 + 3q$

$-4p + 6q = 10$

6. $5x + 9y = -\frac{17}{5}$

$2x - 7y = 5$

7. Two toy robots are placed on a coordinate grid. Both are moving down from the x-axis. One is traveling along the line $x + 2y = \frac{2}{7}$, and the other is traveling along the line $x = -\frac{12}{5}y$. At what point will the robots meet?

Lesson 7.4

NAME _______________________ DATE ___________

Available as a transparency

Application Lesson Opener

For use with pages 418–424

Use the following information for Questions 1 and 2.

You sold two different types of wrapping paper for your band fund-raiser. One type sold for $6 a roll and the other for $8 a roll. You collected a total of $92 for the 14 rolls you sold.

1. Let x represent the number of $6 rolls you sold and y the number of $8 rolls you sold. Which system of equations can be used to model this problem? Why?

A. $x + y = 92$
$6x + 8y = 14$

B. $x + y = 14$
$6x + 8y = 92$

C. $x - y = 14$
$6x - 8y = 92$

D. $x - y = 92$
$6x - 8y = 92$

2. What method would you use to solve the system of equations you chose in Question 1? Explain your answer.

Use the following information for Questions 3 and 4.

You paid $31 to ship 8 packages. The shipping for each package in one group was $3.50. The shipping for each package in the other group was $5.

3. Let x represent the number of $3.50 packages and y represent the number of $5 packages. Which system of equations can be used to model this problem? Why?

A. $x + y = 3.50$
$x + y = 5$

B. $x + y = 31$
$3.5x + 5y = 8$

C. $x + y = 31$
$x - y = 8$

D. $x + y = 8$
$3.5x + 5y = 31$

4. What method would you use to solve the system of equations you chose in Question 3? Explain your answer.

LESSON **7.4**

NAME ______________________________ DATE ____________

Practice A

For use with pages 418–424

Choose the method to solve the linear system. Explain your choice.

1. $2x - 3y = 24$
   $2x + y = 8$
2. $x - y = 4$
   $x + y = 8$
3. $y - 3x = 7$
   $y + 2x = 2$
4. $2x + y = 5$
   $x - y = 1$
5. $3x - y = 9$
   $x + 2y = 10$
6. $x + y = 50$
   $3x - 2y = 0$

Choose a method to solve the linear system. Explain your choice, and then solve the system.

7. $6x + 9y = -6$
   $x + y = 1$
8. $x + 4y = 1$
   $2x + 7y = 3$
9. $3x + 4y = 4$
   $y = x - 6$
10. $4x - 2y = -6$
    $-3x + 2y = -8$
11. $3x + 5y = -13$
    $3x + y = -5$
12. $2x + 3y = -12$
    $2x - 3y = 0$

Solve the linear system using the method of your choice.

13. $4x + 3y = 14$
    $-4x + 5y = 2$
14. $3x + 2y = 13$
    $2x + y = 7$
15. $x - y = 2$
    $3x + y = 10$
16. $x + y = 1$
    $4x - 3y = 18$
17. $4x + 2y = 14$
    $x = 1 + 2y$
18. $y = 4x - 6$
    $3y = 7 - 3x$

Baseball Glove Sales **In Exercises 19 and 20, use the following information.**

A sporting goods store sells right-handed and left-handed baseball gloves. In one month, 12 gloves were sold for a total revenue of \$561. Right-handed gloves cost \$45 and left-handed gloves cost \$52.

19. Let x represents the number of right-handed gloves sold and let y represent the number of left-handed gloves sold. Write a system of equations you could solve to find the number of each type of glove sold.
20. Solve the system.

23. ***Southern Cuisine*** Your family goes to a Southern-style restaurant for dinner. There are 6 people in your family. Some order the chicken dinner for \$14 and some order the steak dinner for \$17. If the total bill was \$99 how many people ordered each dinner?

Cookout **In Exercises 21 and 22, use the following information.**

You are buying the meat for a cookout. You need to buy 8 packages of meat. A package of hotdogs costs \$1.60 and a package of hamburgers costs \$5. You spend a total of \$23.

21. Let x represent the number of packages of hotdogs bought and let y represent the number of packages of hamburgers bought. Write a system of equations you could solve to find the number of packages of each type of meat bought.
22. Solve the system.

24. ***Dimensions of a Rectangle*** The perimeter of the rectangle is 20 inches. The perimeter of the inscribed triangle is 20 inches. Find the dimensions of the rectangle.

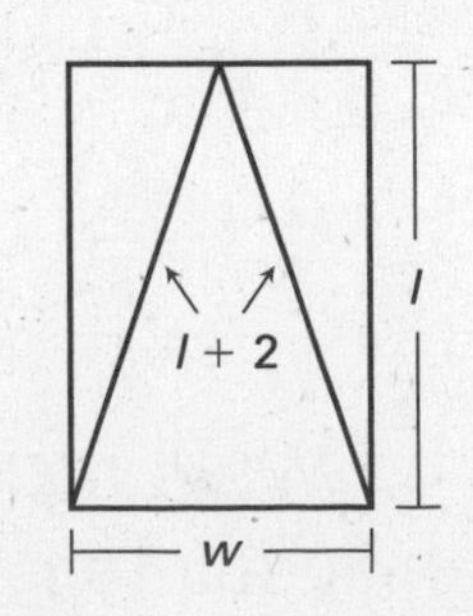

LESSON **7.4**

NAME ______________________ DATE __________

Practice B

For use with pages 418–424

Choose a method to solve the linear system. Explain your choice, and then solve the system.

1. $2x - 3y = 24$
$2x + y = 8$

2. $x - y = 4$
$x + y = 8$

3. $y - 3x = 7$
$y + 2x = 2$

4. $2x + y = 5$
$x - y = 1$

5. $3x - y = 9$
$x + 2y = 10$

6. $x + y = 50$
$3x - 2y = 0$

Solve the linear system using the method of your choice.

7. $6x + 9y = 3$
$x + 4y = -2$

8. $-x = 10$
$2x + 7y = 1$

9. $-3x + y = -4$
$y = x - 6$

10. $4x - 6 = 2y$
$-3x + 2y = -3$

11. $-3x + 5y = -10$
$-3x + 6y = -12$

12. $2x + 3y = 8$
$2x - 3y = -4$

13. $4x - 3y = -4$
$-3x + 5y = -8$

14. $1.8x + 3y = 3$
$-2x - 2.5y = -5$

15. $x - y = 2$
$3x + y = -10$

16. $2x + 4y = -1$
$4x - 3y = -2$

17. $6x - 3y = -5$
$x - \frac{2}{3}y = -1$

18. $y = \frac{1}{2}x - 4$
$x = -2 + \frac{1}{3}y$

Cookout **In Exercises 19 and 20, use the following information.**

You are buying the meat for a cookout. You need to buy 8 packages of meat. A package of hotdogs costs \$1.89 and a package of hamburgers costs \$5.19. You spend a total of \$31.62.

19. Let x represent the number of packages of hotdogs bought and let y represent the number of packages of hamburgers bought. Write a system of equations you could solve to find the number of packages of each type of meat bought.

20. Solve the system.

21. ***Baseball Glove Sales*** A sporting goods store sells right-handed and left-handed baseball gloves. In one month, 12 gloves were sold for a total revenue of \$561. Right-handed gloves cost \$45 and left-handed gloves cost \$52. Find the number of each type of glove sold.

22. ***Southern Cuisine*** Your family goes to a Southern-sytle restaurant for dinner. There are 6 people in your family. Some order the chicken dinner for \$14.80 and some order the steak dinner for \$17. If the total bill was \$91, how many people ordered each dinner?

23. ***Dimensions of a Rectangle*** The perimeter of the rectangle is 21 inches. The perimeter of the inscribed triangle is 21 inches. Find the dimensions of the rectangle.

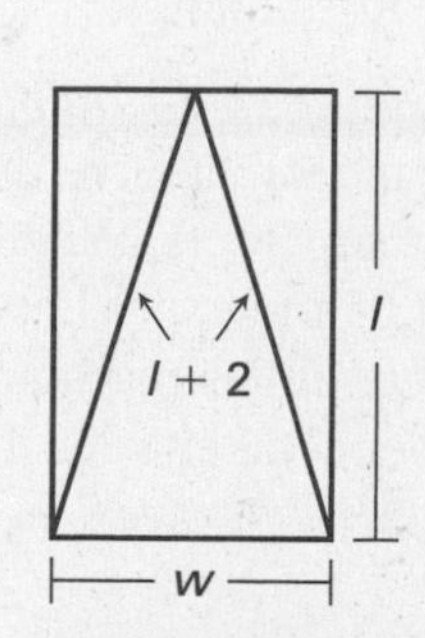

NAME ______________________________ DATE __________

Practice C

For use with pages 418–424

Choose a method to solve the linear system. Explain your choice, and then solve the system.

1. $4x - 3y = -20$
   $4x + 2y = 0$
2. $5x + 2y = -1$
   $-3x + 4y = -15$
3. $-y = -6$
   $x + 2y = 8$
4. $3y = 7x - 5$
   $13 - 3y = 2x$
5. $-3x - y = 9$
   $y + x = -3$
6. $1.2x - 2.5y = -3.5$
   $1.4x - 1.5y = -5.5$

Solve the linear system using the method of your choice.

7. $7x + 9y = 5$
   $x + 5y = -3$
8. $-x = 11$
   $2x + 7y = -1$
9. $-4x + y = -3$
   $y = x - 6$
10. $4x - 3 = 3y$
    $-5x + 3y = -6$
11. $-3x + 8y = -16$
    $-3x + 6y = -12$
12. $2x + 3y = 13$
    $2x - 3y = -5$
13. $4x - 3y = -6$
    $-3x + 5y = -12$
14. $2y + 3x = 10$
    $2x = 5 - 3y$
15. $x = y + 2$
    $5x + y = -20$
16. $-2x + 4y = 1$
    $-8x - 3y = 4$
17. $1.8x + 4y = -1$
    $-2x - 3.5y = 3$
18. $y = \frac{1}{2}x - 6$
    $x = \frac{2}{3}y$
19. $7x - 14y = 2$
    $-\frac{3}{5}x + y = 0$
20. $-0.7x + 0.1y = 0.9$
    $x - 0.3y = 1.7$
21. $\frac{1}{2}x + \frac{3}{4}y = 9$
    $-2x + y = -4$

Cookout **In Exercises 22 and 23, use the following information.**

You are buying the meat for a cookout. You need to buy 8 packages of meat. A package of hotdogs costs $2.19 and a package of hamburgers costs $5.89. You spend a total of $28.62.

22. Write a system of equations you could solve to find the number of packages of each type of meat bought.
23. Solve the system.
24. ***Southern Cuisine*** Your family goes to a Southern-style restaurant for dinner. There are 6 people in your family. Some order the chicken dinner for $14.89 and some order the steak dinner for $17.69. If the total bill was $100.54, how many people ordered each dinner?
25. ***Dimensions of a Rectangle*** The perimeter of the rectangle is 22 inches. The perimeter of the inscribed triangle is 21.5 inches. Find the dimensions of the rectangle.

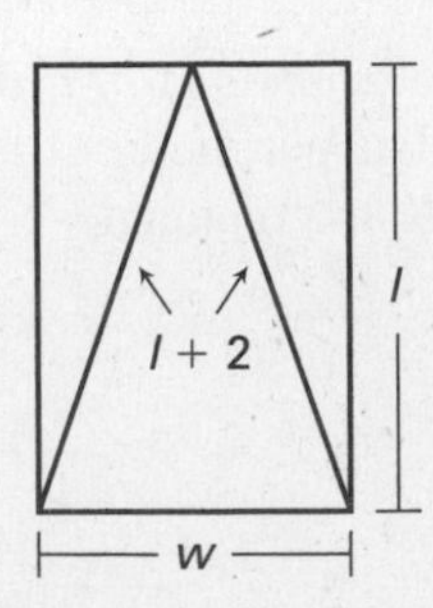

26. ***Driving to Grandma's House*** You live in Pennsylvania and your grandparents live in Ohio. Your family decides to take a trip to visit your grandparents. When you are in Pennsylvania, you drive at an average rate of 55 miles per hour. When you are in Ohio, you drive at an average rate of 65 miles per hour. The entire trip of 295 miles takes 5 hours. How long does it take to reach the Pennsylvania-Ohio border? How long does it take to get from the border to your grandparent's house?

LESSON 7.4

NAME ______________________ DATE ________

Reteaching with Practice

For use with pages 418–424

GOAL Choose the best method to solve a linear system and use a system to model real-life problems

EXAMPLE 1 *Choosing a Solution Method*

Your cousin borrowed \$6000, some on a home-equity loan at an interest rate of 9.5% and the rest on a consumer loan at an interest rate of 11%. Her total interest paid was \$645. How much did she borrow at each rate?

SOLUTION

Verbal Model

[Home-equity loan amount] + [Consumer loan amount] = [Total loan]

[Home equity loan rate] · [Home-equity loan amount] + [Consumer loan rate] · [Consumer loan amount] = [Total interest paid]

Labels

Home-equity loan amount $= x$		(dollars)
Consumer loan amount $= y$		(dollars)
Total loan $= 6000$		(dollars)
Home-equity loan rate $= 0.095$		(percent written in decimal form)
Consumer loan rate $= 0.11$		(percent written in decimal form)
Total interest paid $= 645$		(dollars)

Algebraic Model

$x + y = 6000$ Equation 1 (loan)

$0.095x + 0.11y = 645$ Equation 2 (interest)

Because the coefficients of x and y are 1 in Equation 1, use the substitution method. You can solve Equation 1 for x and substitute the result into Equation 2. You will obtain 5000 for y. Substitute 5000 into Equation 1 and solve for x. You will obtain 1000 for x.

The solution is \$1000 borrowed at 9.5% and \$5000 borrowed at 11%.

Exercise for Example 1

1. Choose a method to solve the linear system. Explain your choice.

a. $2x - y = 3$
$x + 3y = 5$

b. $4x + 4y = 16$
$-2x + 5y = 9$

c. $x - 3y = 3$
$5x + 2y = 14$

NAME ______________________ DATE __________

Reteaching with Practice

For use with pages 418–424

EXAMPLE 2 Solving a Cost Problem

For a community bake sale, you purchased 12 pounds of sugar and 15 pounds of flour. Your total cost was $9.30. The next day, at the same prices, you purchased 4 pounds of sugar and 10 pounds of flour. Your total cost the second day was $4.60. Find the cost per pound of the sugar and the flour purchases.

SOLUTION

Verbal Model

Amount of sugar Day 1	·	Cost of sugar	+	Amount of flour Day 1	·	Cost of flour	=	Total cost Day 1
Amount of sugar Day 2	·	Cost of sugar	+	Amount of flour Day 2	·	Cost of flour	=	Total cost Day 2

Labels

Amount of sugar Day 1 = 12	(pounds)
Amount of flour Day 1 = 15	(pounds)
Amount of sugar Day 2 = 4	(pounds)
Amount of flour Day 2 = 10	(pounds)
Cost of sugar = x	(dollars per pound)
Cost of flour = y	(dollars per pound)
Total cost Day 1 = 9.30	(dollars)
Total cost Day 2 = 4.60	(dollars)

Algebraic Model

$12x + 15y = 9.30$ Equation 1 (Purchases–Day 1)

$4x + 10y = 4.60$ Equation 2 (Purchases–Day 2)

Use linear combinations to solve this linear system because none of the variables has a coefficient of 1 or -1. You can get the coefficients of x to be opposites by multiplying Equation 2 by -3. You will obtain 0.30 for y. Substitute 0.30 for y into Equation 1 and solve for x. You will obtain 0.40 for x.

The solution of the linear system is (0.40, 0.30). You conclude that sugar costs $.40 per pound and flour costs $.30 per pound.

Exercise for Example 2

2. Rework Example 2 if the cost of the first purchase was $7.95 and the cost of the second purchase was $3.90.

LESSON 7.4

NAME ______________________________ DATE __________

Quick Catch-Up for Absent Students

For use with pages 418–424

The items checked below were covered in class on (date missed) __________

Lesson 7.4: Applications of Linear Systems

____ **Goal 1:** Choose the best method to solve a system of linear equations. (p. 418)

Material Covered:

____ Student Help: Study Tip

____ Example 1: Choosing a Solution Method

____ **Goal 2:** Use a system to model real-life problems. (pp. 419–420)

Material Covered:

____ Example 2: Solving a Mixture Problem

____ Example 3: Making a Decision

____ Other (specify) ______________________________

____

Homework and Additional Learning Support

____ Textbook (specify) pp. 421–424

____

____ *Reteaching with Practice* worksheet (specify exercises) __________

____ *Personal Student Tutor* for Lesson 7.4

Lesson 7.4

NAME ______________________ DATE __________

Cooperative Learning Activity

For use with pages 418–424

GOAL **To compare music prices using systems of equations**

Materials: paper, graph paper, pencil

Exploring Systems of Linear Equations

Todd purchases all of his CDs in a local store. All CDs in this store cost \$13.99. Todd's sister Shelli has just decided to join a mail-order music club. Her first ten CDs are free. After that, each CD will cost her \$16.99 (including shipping charges). Use this information for the activity below.

Instructions

1. Write an equation that models the cost of Todd's CD purchases and the cost of Shelli's CD purchases. Let y be the total cost of the CDs, and let x be the number of CDs purchased.
2. Use the substitution method to solve this system of equations.
3. Verify your answer by solving the system of equations using the combination method.
4. Finally, graph the two equations.

Analyzing the Results

1. When will Todd and Shelli have spent the same amount for CDs? How many CDs will they have purchased?

2. When would it be less expensive to buy CDs from the music store? When would it be less expensive to buy CDs through the mail?

LESSON 7.4

NAME ______________________________ DATE ____________

Interdisciplinary Application

For use with pages 418–424

Brass Instruments

MUSIC Wind instruments are played by blowing into or through a tube. There are two chief types. They are woodwind instruments and brass instruments.

All woodwind instruments except the saxophone at one time were made of wood. Today, many are made of metal or other materials. In such woodwinds as recorders, the player blows through a mouthpiece into the instrument. In some other woodwinds, such as flutes and piccolos, the player blows across a hole in the instrument. Still other woodwinds, called reed instruments, have one or two reeds attached to the mouthpiece. The reeds vibrate when the musician blows on them. The player controls the pitch by placing the fingers on holes in the instrument or on keys that cover holes. In this way, the player lengthens or shortens the column of air that vibrates inside the instrument.

Brass instruments are played differently from woodwind instruments. The player presses the lips against the instrument's mouthpiece so that the lips vibrate like reeds when the player blows. By either tensing or relaxing the lips, the player produces different pitches. With most brass instruments, the player can further control the pitch with valves that lengthen or shorten the tube through which the air is blown.

In Exercises 1-5, use the following information.

You are in the band room when a shipment of new trumpets and trombones arrives. Your music teacher asks you to make sure the order is right. All the boxes are about the same size, so you find the number of boxes in the shipment. There are 27 total boxes. The bill shows the school paid $10,950. In the catalogue, the price of a new trumpet is $350 and the price of a new trombone is $475.

1. Write an equation representing the total number of instruments.
2. Write an equation representing the total cost.
3. Using your equations from Exercises 1 and 2, write and solve the linear system.
4. Choose a different method from the method you used in Exercise 3 to check your results.
5. How many trumpets did your school receive? How many trombones?

NAME ______________________________ DATE __________

Challenge: Skills and Applications

For use with pages 418–424

In Exercises 1–2, choose a method to solve the linear system. Explain your choice, and then solve the system.

1. $3x + y = a$
$x - 2y = 5a$

2. $2ax + by = 10$
$ax - 2by = 20$

3. When Lauren Bauer received her bank statement, she noticed that she had recorded the wrong amount for one check by reversing the tens' digit and the ones' digit, causing a mistake of $36. Her friend, Olga Sven, told her that reversing the digits always causes a mistake that is a multiple of 9. Was Olga correct? To find out, let u be the ones' digit and let d be the tens' digit of the original amount of the check, and let p and q be the original and mistaken amounts, respectively. What do you find when you subtract the two resulting equations?

In Exercises 4–6, use the following information.

Yoshi Tanaka kept track of his car's mileage and found that on 9 gallons of gasoline he was able to drive a combination of 96 miles in the city and 160 miles on the highway. On 11 gallons of gasoline, he was able to drive a combination of 144 miles in the city and 160 miles on the highway. Yoshi wants to know how many miles per gallon his car gets in city driving and how many it gets in highway driving.

4. Write a linear system to model this situation, where x represents the miles per gallon in the city and y represents the miles per gallon in highway driving.

5. Let $u = \frac{1}{x}$ and $v = \frac{1}{y}$. Substitute and solve for u and v.

6. Substitute for u in $u = \frac{1}{x}$ and solve for x. Similarily, substitute and solve for y in $v = \frac{1}{y}$. How many miles per gallon does Yoshi's car get in city driving and how many does it get in highway driving?

TEACHER'S NAME ______________________ CLASS ______ ROOM ______ DATE ______

Lesson Plan

1-day lesson (See *Pacing the Chapter,* TE pages 394C–394D) **For use with pages 425–431**

1. **Identify linear systems as having one solution, no solution, or infinitely many solutions.**
2. **Model real-life problems using a linear system.**

State/Local Objectives ____

____

✓ Check the items you wish to use for this lesson.

STARTING OPTIONS

____ Homework Check: TE page 421; Answer Transparencies
____ Warm-Up or Daily Homework Quiz: TE pages 426 and 424, CRB page 65, or Transparencies

TEACHING OPTIONS

____ Motivating the Lesson: TE page 427
____ Concept Activity: SE page 425
____ Lesson Opener (Graphing Calculator): CRB page 66 or Transparencies
____ Graphing Calculator Activity with Keystrokes: CRB pages 67–69
____ Examples 1–4: SE pages 426–428
____ Extra Examples: TE pages 427–428 or Transparencies
____ Closure Question: TE page 428
____ Guided Practice Exercises: SE page 429

APPLY/HOMEWORK

Homework Assignment

____ Basic 12–17, 18–28 even, 31, 32, 36–38, 43–48
____ Average 12–17, 18–28 even, 30–33, 36–38, 43–48
____ Advanced 12–17, 18–28 even, 30–33, 36–48

Reteaching the Lesson

____ Practice Masters: CRB pages 70–72 (Level A, Level B, Level C)
____ Reteaching with Practice: CRB pages 73–74 or Practice Workbook with Examples
____ Personal Student Tutor

Extending the Lesson

____ Applications (Real-Life): CRB page 76
____ Challenge: SE page 431; CRB page 77 or Internet

ASSESSMENT OPTIONS

____ Checkpoint Exercises: TE pages 427–428 or Transparencies
____ Daily Homework Quiz (7.5): TE page 431, CRB page 80, or Transparencies
____ Standardized Test Practice: SE page 431; TE page 431; STP Workbook; Transparencies

Notes ____

____

____

LESSON **7.5**

TEACHER'S NAME ____________________ CLASS ______ ROOM ______ DATE ______

Lesson Plan for Block Scheduling

Half-day lesson (See *Pacing the Chapter,* TE pages 394C–394D) **For use with pages 425–431**

1. Identify linear systems as having one solution, no solution, or infinitely many solutions.
2. Model real-life problems using a linear system.

State/Local Objectives ______________________________

____

____

CHAPTER PACING GUIDE	
Day	**Lesson**
1	7.1 (all); 7.2 (all)
2	7.3 (all)
3	7.4 (all)
4	**7.5 (all)**; 7.6 (begin)
5	7.6 (end); Review Ch. 7
6	Assess Ch. 7; 8.1 (begin)

✓ Check the items you wish to use for this lesson.

STARTING OPTIONS

____ Homework Check: TE page 421; Answer Transparencies
____ Warm-Up or Daily Homework Quiz: TE pages 426 and 424, CRB page 65, or Transparencies

TEACHING OPTIONS

____ Motivating the Lesson: TE page 427
____ Concept Activity: SE page 425
____ Lesson Opener (Graphing Calculator): CRB page 66 or Transparencies
____ Graphing Calculator Activity with Keystrokes: CRB pages 67–69
____ Examples 1–4: SE pages 426–428
____ Extra Examples: TE pages 427–428 or Transparencies
____ Closure Question: TE page 428
____ Guided Practice Exercises: SE page 429

APPLY/HOMEWORK

Homework Assignment (See also the assignment for Lesson 7.6.)

____ Block Schedule: 12–17, 18–28 even, 30–33, 36–38, 43–48

Reteaching the Lesson

____ Practice Masters: CRB pages 70–72 (Level A, Level B, Level C)
____ Reteaching with Practice: CRB pages 73–74 or Practice Workbook with Examples
____ Personal Student Tutor

Extending the Lesson

____ Applications (Real-Life): CRB page 76
____ Challenge: SE page 431; CRB page 77 or Internet

ASSESSMENT OPTIONS

____ Checkpoint Exercises: TE pages 427–428 or Transparencies
____ Daily Homework Quiz (7.5): TE page 431, CRB page 80, or Transparencies
____ Standardized Test Practice: SE page 431; TE page 431; STP Workbook; Transparencies

Notes ______________________________

____

____

NAME ______________________ DATE __________

Available as a transparency

Warm-Up Exercises

For use before Lesson 7.5, pages 425–431

Write each equation in slope-intercept form. Then identify the slope and *y*-intercept.

1. $x + y = 12$

2. $-x + 3y = 3$

3. $4x - 6y = 15$

4. $x - 5y = 10$

Daily Homework Quiz

For use after Lesson 7.4, pages 418–424

Choose a method to solve the linear system.

1. $x - 2y = 5$
$x + 3y = -10$

2. $3x + 7y = -29$
$4x + 3y = -7$

3. $6x - 5y = 0$
$3x + 2y = 9$

4. $4x + y = 7$
$9x + 2y = 16$

5. $y = 1 - x$
$4x - 2y = -1$

6. $5x + 3y = 9$
$-2x - 5y = 23$

7. Music club A has a \$30 membership fee and charges \$8 per CD. Music club B has a \$45 membership fee and charges \$7 per CD. How many CDs would a person need to buy for club A to be the more expensive?

LESSON 7.5

NAME ______________________ DATE ____________

Available as a transparency

Graphing Calculator Lesson Opener

For use with pages 425–431

Lesson 7.5

Enter the linear system into your calculator. Describe the two lines graphed and the solution.

1. $x + y = 3$
   $2x + y = 2$

2. $x - 2y = -6$
   $-2x + 4y = 12$

3. $3x + y = 1$
   $3x + y = -2$

4. $x - y = -2$
   $x - y = 3$

5. $2x - y = 3$
   $3x + y = 1$

6. $x - 3y = 2$
   $3y - x = -2$

7. $2x - 4y = 6$
   $x - 2y = 3$

8. $2x - y = 3$
   $3x + y = 1$

9. Which systems have one solution? What is true about the graphs of these linear systems?

10. Which systems have no solutions? What is true about the graphs of these linear systems?

11. Which systems have infinitely many solutions? What is true about the graphs of these linear systems?

12. Make a conjecture about the graph of a linear system and its solution.

NAME ______________________ DATE __________

Graphing Calculator Activity

For use with pages 426–431

GOAL **To discover how to shift the graph of a square root function horizontally and vertically**

The standard form of the square root function is $y = a\sqrt{x - h} + k$. In the function $y = \sqrt{x}$, which was presented in Lesson 7.4, you do not see the a, h, or k because $a = 1$ and both h and k are zero.

Activity

1. Enter the equation $y = \sqrt{x}$ into your graphing calculator as Y_1 and plot the graph in a standard viewing window.
2. Enter and plot the graph of each equation one at a time as Y_2. Compare each graph with the graph of $y = \sqrt{x}$ from Step 1.

   a. $y = \sqrt{x - 5}$ b. $y = \sqrt{x + 7}$
3. Identify h in each equation of Step 2.
4. Enter and plot the graph of each equation one at a time as Y_2. Compare each graph with the graph of $y = \sqrt{x}$ from Step 1.
5. Identify k in each equation of Step 4.
6. Enter $y = \sqrt{x - 2} + 8$ as Y_2 and plot the graph. Compare the graph with the graph of $y = \sqrt{x}$ from Step 1.

Exercises

1. Sketch the graph of each equation below. Use your graphing calculator to check your answer.

   a. $y = \sqrt{x - 3} + 5$ b. $y = \sqrt{x + 1} - 4$ c. $y = \sqrt{x + 2} + 6$

In Exercises 2–5, complete the statement.

2. If h is positive, the graph will shift to the ________ h units.
3. If h is negative, the graph will shift to the ________ h units.
4. If k is positive, the graph will shift ________ k units.
5. If k is negative, the graph will shift ________ k units.

See page 68 for keystrokes.

NAME ______________________ DATE __________

Graphing Calculator Activity

For use with pages 426–431

TI-82

[Y=] [2nd] [√] [X,T,θ] [ENTER]
[ZOOM] 6
[Y=] [ENTER] [2nd] [√] [(] [X,T,θ] [−]
5 [)] [ENTER] [GRAPH]
[Y=] [ENTER] [CLEAR] [2nd] [√] [(]
[X,T,θ] [+] 7 [)] [ENTER]
[GRAPH]
[Y=] [ENTER] [CLEAR] [2nd] [√]
[X,T,θ] [+] 6 [ENTER]
[GRAPH]
[Y=] [ENTER] [CLEAR] [2nd] [√]
[X,T,θ] [−] 4 [ENTER]
[GRAPH]
[Y=] [ENTER] [2nd] [√] [(] [X,T,θ] [−]
2 [)] [+] 8 [ENTER]
[GRAPH]

TI-83

[Y=] [2nd] [√] [X/θ/T/n] [)] [ENTER]
[ZOOM] 6
[Y=] [ENTER] [2nd] [√] [X,T,θ,n] [−] 5
[)] [ENTER] [GRAPH]
[Y=] [ENTER] [CLEAR] [2nd] [√]
[X,T,θ,n] [+] 7 [)] [ENTER]
[GRAPH]
[Y=] [ENTER] [CLEAR] [2nd] [√] [X,T,θ,n] [−] 2
[)] [+] 8 [ENTER]
[GRAPH]

SHARP EL-9600c

[Y=] [2ndF] [√] [X/θ/T/n] [ENTER]
[ZOOM] [A] 5
[Y=] [ENTER] [2ndF] [√] [X/θ/T/n] [−] 5 [ENTER]
[GRAPH]
[Y=] [ENTER] [CL] [2ndF] [√]
[X/θ/T/n] [+] 7 [ENTER] [GRAPH]
[Y=] [ENTER] [CL] [2ndF] [√]
[X/θ/T/n] [▶] [÷] 6 [ENTER]
[GRAPH]
[Y=] [ENTER] [CL] [2ndF] [√]
[X/θ/T/n] [▶] [−] 6 [ENTER]
[GRAPH]
[Y=] [ENTER] [CL] [2ndF] [√] [X/θ/T/n] [−]
2 [▶] [+] 8 [ENTER]
[GRAPH]

CASIO CFX-9850GA PLUS

From the main menu, choose GRAPH.

[SHIFT] [√] [X,θ,T] [EXE]
[SHIFT] [F3] [F3] [EXIT] [F6]
[EXIT] [SHIFT] [√] [(] [X,θ,T] [−]
5 [)] [EXE] [F6]
[EXIT] [▲] [SHIFT] [√] [(] [X,θ,T] [−]
7 [)] [EXE] [F6]
[EXIT] [▲] [SHIFT] [√] [X,θ,T] [+] 6
[EXE] [F6]
[EXIT] [▲] [SHIFT] [√] [X,θ,T] [−] 4
[EXE] [F6]
[EXIT] [▲] [SHIFT] [√] [(] [X,θ,T] [−]
2 [)] [+] 8 [EXE] [F6]

NAME ______________________ DATE ____________

Graphing Calculator Activity Keystrokes

For use with pages 430–436

Excel Keystrokes for Exercise 34

Open computer to excel program.

Select cell A1.

x TAB $y = 2x + 3$ TAB $y = 2x - 9$ TAB col. B–col. C ENTER

Enter x-values -3 to 4 in cells A2–A9.

Select cell B2.

= 2*A2 + 3 ENTER

Select cell B2. From the **Edit** menu, choose **Copy**.

Select cells B3–B9. from the **Edit** menu choose **Paste.**

Select cell C2.

= 2*A2 − 9 ENTER

Select cell C2. From the **Edit** menu, choose **Copy.**

Select cells C3–C9. from the **Edit** menu, choose **Paste.**

Select cell D2.

= B2 − C2 ENTER

Select cell D2. From the **Edit** menu, choose **Copy.**

Select cells D3–D9. From the **Edit** menu, choose **Paste.**

NAME ______________________________ DATE __________

Practice A

For use with pages 426–431

Match the graph with its linear system. Does the system have exactly one solution, no solution, or infinitely many solutions?

A. $-2x + y = 1$
$-4x + 2y = -6$

B. $x - y = 4$
$x + y = 4$

C. $6x + 3y = 6$
$2x + y = 2$

D. $2x + y = 2$
$-2x - y = 0$

E. $-2x + y = -4$
$2x + y = 2$

F. $x - y = 2$
$3x - 3y = 6$

1.
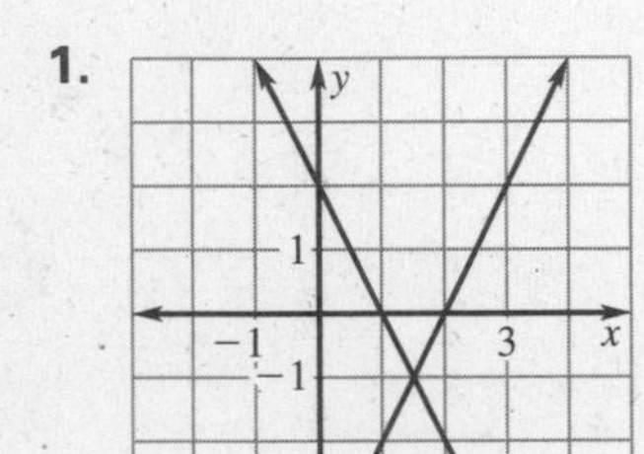

2.
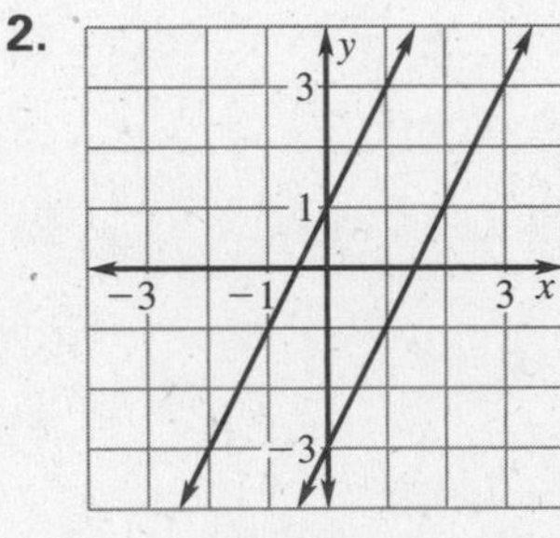

3.
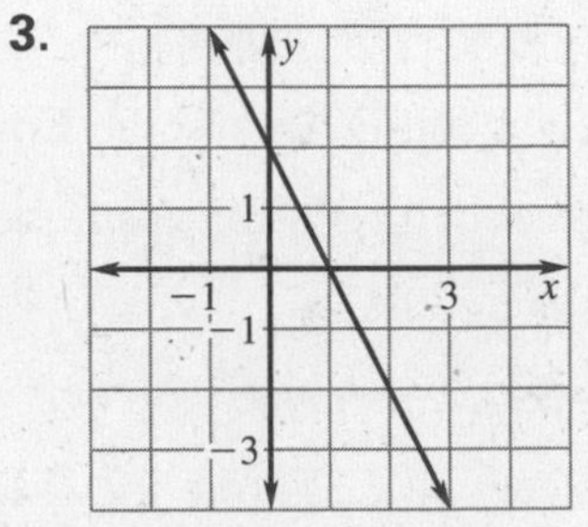

4.
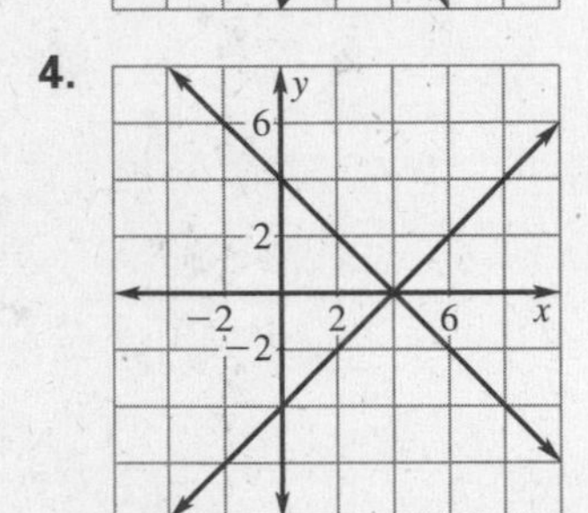

5.
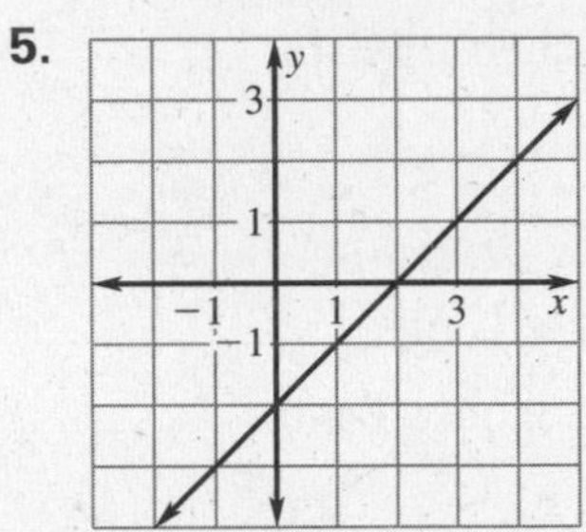

6.
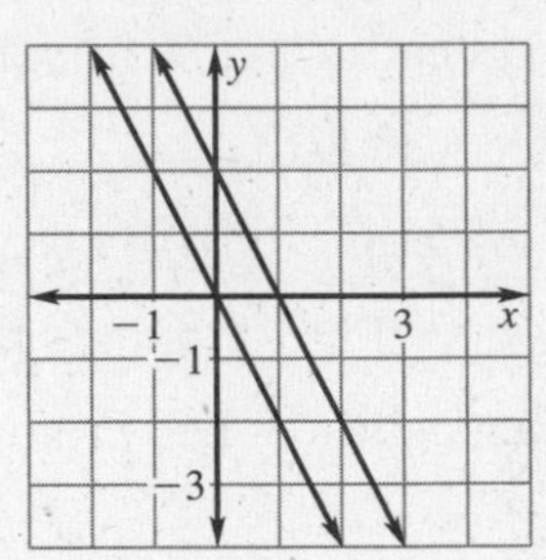

Use the substitution method or linear combinations to solve the linear system and tell how many solutions the system has.

7. $x + y = -1$
$x + y = 8$

8. $x - 3y = 2$
$-2x + 6y = 2$

9. $3x - 2y = 0$
$3x - 2y = -4$

10. $6x + 4y = 14$
$3x + 2y = 2$

11. $3x - 2y = 3$
$-6x + 4y = -6$

12. $-2x + 4y = -2$
$-x - 2y = 3$

Use the graphing method to solve the linear system and tell how many solutions the system has.

13. $x - y = 5$
$x - y = 2$

14. $3x - 10y = -15$
$-3x + 10y = 15$

15. $-x + 4y = -1$
$3x - 12y = -3$

16. $6x - 5y = 3$
$-12x + 10y = 5$

17. $-3x - 2y = 6$
$-6x + 4y = -12$

18. $6x - 3y = 4$
$-4x + 2y = -\frac{8}{3}$

19. *U.S. Population* The male and female populations of the United States from 1960 to 1990 are shown in the matrix. Construct two scatter plots, one for the male population and one for the female population. Then find the line that best fits each scatter plot.

20. Discuss the linear system you found in Exercise 19. Are the two lines parallel? Do you think that the number of men in the U.S. will equal the number of women before the year 2000? Explain.

Population (in millions)

	Male	Female
1960	89	91
1970	100	105
1980	111	117
1990	122	128

NAME ______________________________ DATE __________

Practice B

For use with pages 426–431

Match the graph with its linear system. Does the system have exactly one solution, no solution, or infinitely many solutions?

A. $-2x + y = 6$
$-4x + 2y = -6$

B. $x - 4y = 7$
$5x + y = -7$

C. $-9x + 3y = -6$
$-3x + y = -2$

D. $5x + 4y = 2$
$-5x - 4y = -1$

E. $-2x + 3y = -6$
$2x + 3y = 0$

F. $x - y = 2$
$7x - 7y = 14$

1.

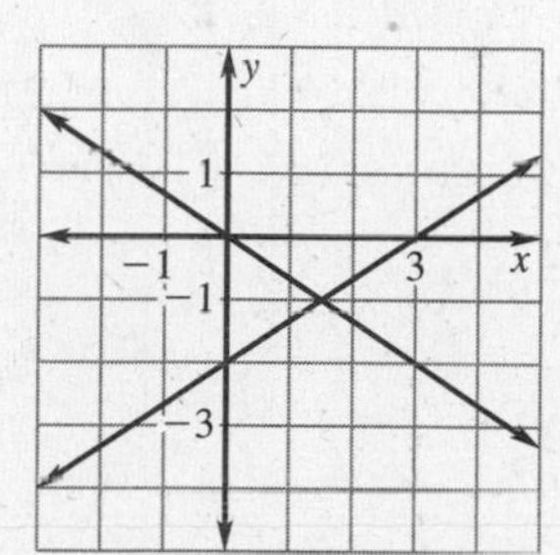

2.

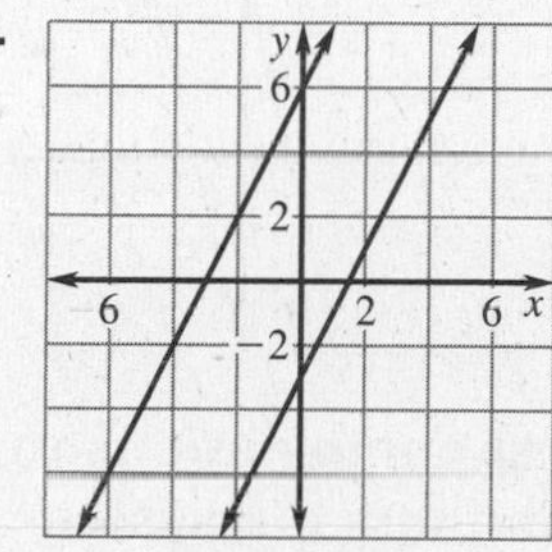

3.

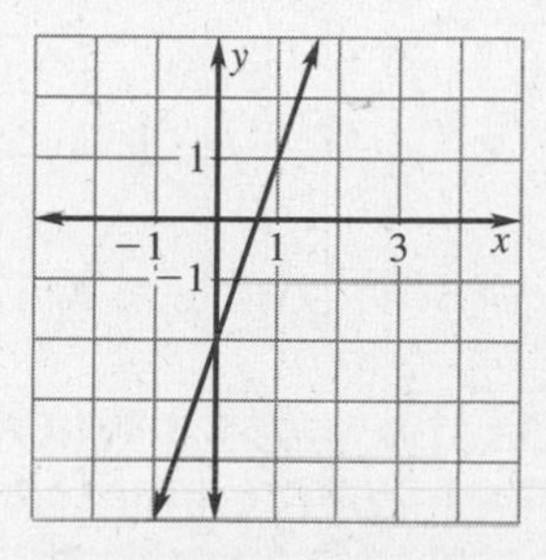

4.

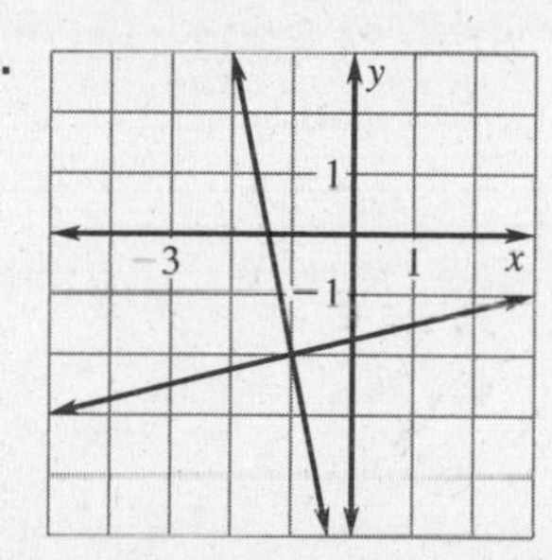

5.

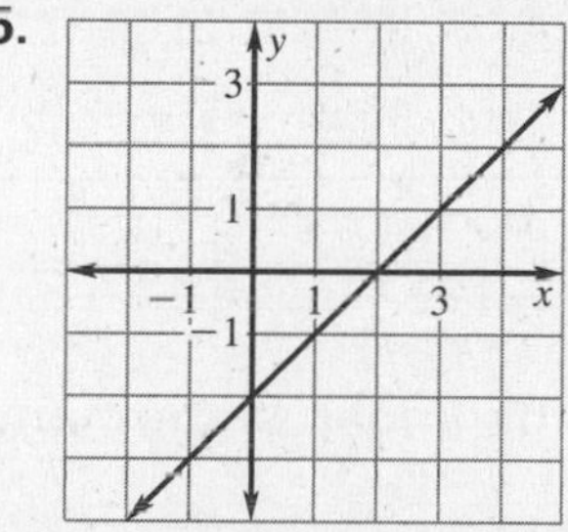

6.

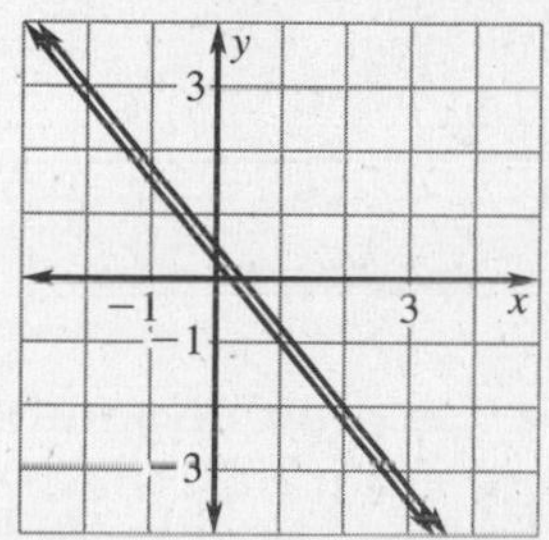

Use the substitution method or linear combinations to solve the linear system and tell how many solutions the system has.

7. $-8x + 8y = -6$
$3x - 3y = 8$

8. $-6x - 6y = -12$
$-2x - 2y = -4$

9. $-4x - 2y = 2$
$4x - 2y = 18$

10. $6x - 4y = -6$
$3x + 2y = 1$

11. $3x - 2y = -5$
$-9x + 6y = 15$

12. $x + 3y = -3$
$\frac{1}{3}x + y = 1$

Use the graphing method to solve the linear system and tell how many solutions the system has.

13. $2x + y = 7$
$4x + 2y = -10$

14. $-2x + 3y = 18$
$-2x + 3y = -18$

15. $-x + 4y = -3$
$3x - 12y = 3$

16. $6x - 5y = 3$
$-2x + \frac{5}{3}y = 1$

17. $x - 7y = 10$
$-6x + 4y = -22$

18. $\frac{1}{2}x + y = -2$
$\frac{3}{2}x + 3y = 6$

19. ***Revenue and Cost*** The matrix gives the revenue and cost of running a business from 1997 to 2000. Construct two scatter plots, one for revenue and one for cost. Then find the line that best fits each scatter plot.

20. ***Profit*** Profit can be defined as revenue minus cost. What does the graph from Exercise 19 tell you about the business' profit from 1997 to 2000?

Amount (in $1000)

	Revenue	Cost
1997	50	25
1998	100	75
1999	150	125
2000	200	175

LESSON **7.5**

NAME ______________________________ DATE __________

Practice C

For use with pages 426–431

Match the graph with its linear system. Does the system have exactly one solution, no solution, or infinitely many solutions?

A. $-4x + y = 3$
$-8x + 2y = -6$

B. $-2x + y = 1$
$2x + y = 2$

C. $-6x + 3y = -9$
$-4x + 2y = -6$

1.

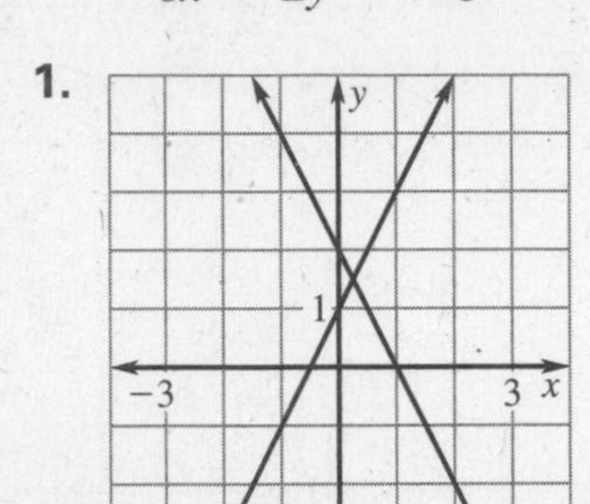

2.

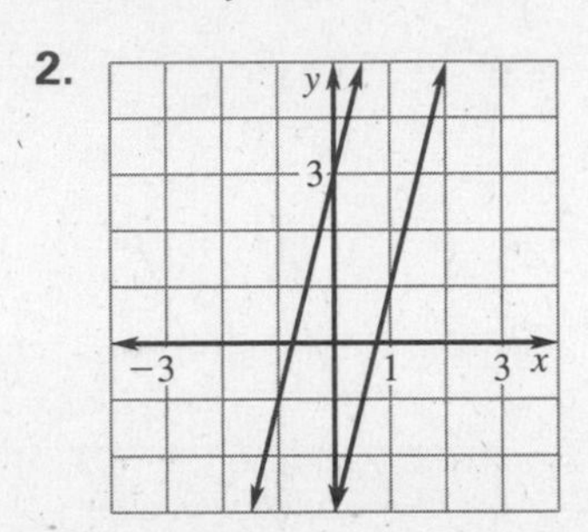

3.

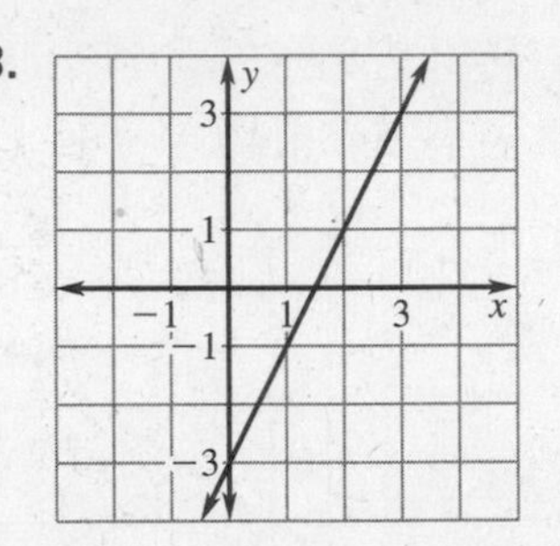

Use the substitution method or linear combinations to solve the linear system and tell how many solutions the system has.

4. $-8x + 8y = -16$
$5x - 5y = 8$

5. $6x - 6y = -14$
$3x - 3y = -7$

6. $3x - 2y = 0$
$\frac{3}{2}x - y = 0$

7. $-5x + 4y = 1$
$4x - 5y = 1$

8. $3x - y = -2$
$-15x + 5y = 0$

9. $-2x + 4y = 1$
$\frac{3}{2}x - 3y = \frac{3}{4}$

Use the graphing method to solve the linear system and tell how many solutions the system has.

10. $2x - 6y = 5$
$3x - 9y = 2$

11. $-2x + 5y = -18$
$-2x + 5y = 18$

12. $2x - y = 3$
$x - \frac{1}{2}y = \frac{3}{2}$

13. $8x - 5y = 3$
$-2x + \frac{5}{4}y = \frac{3}{4}$

14. $-3x + 4y = -8$
$-4x - 3y = 6$

15. $\frac{1}{2}x + y = -\frac{2}{3}$
$\frac{3}{2}x + 3y = -2$

16. ***Revenue and Cost*** The matrix gives the revenue and cost of running a business from 1997 to 2000. Construct two scatter plots, one for revenue and one for cost. Then find the line that best fits each scatter plot.

Amount (in $1000)

	Revenue	Cost
1997	58	33
1998	105	80
1999	154	129
2000	209	184

17. ***Profit*** Profit can be defined as revenue minus cost. What does the graph from Exercise 16 tell you about the business' profit from 1997 to 2000?

18. ***Traveling Time*** You pick up your mother at work and then drive to your sister's out-of-town soccer game. Your total trip takes 2 hours to drive 110 miles at an average rate of 55 miles per hour. Can you determine how long it takes to get to your mother's office or how much longer it takes to get to the soccer field from her office? If yes, solve. If not, explain why? Use the verbal model to help answer the question.

[Time from home to mother's office] + [Time from office to soccer field] = [Total trip time]

[Average rate] · [Time from home to mother's office] + [Average rate] · [Time from office to soccer field] = [Total distance]

LESSON **7.5**

NAME ______________________________ DATE ____________

Reteaching with Practice

For use with pages 426–431

GOAL **Identify linear systems as having one solution, no solution, or infinitely many solutions and model real-life problems using a linear system**

EXAMPLE 1 *A Linear System with No Solution*

Show that the linear system has no solution.

$3x - y = 1$ Equation 1

$3x - y = -2$ Equation 2

SOLUTION

Method 1: GRAPHING Rewrite each equation in slope-intercept form. Then graph the linear system.

$y = 3x - 1$ Revised Equation 1

$y = 3x + 2$ Revised Equation 2

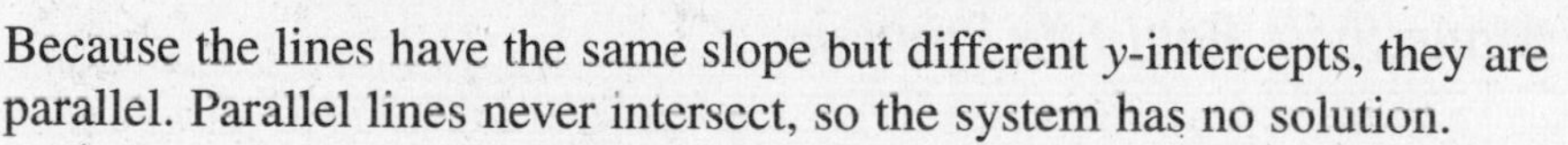

Because the lines have the same slope but different y-intercepts, they are parallel. Parallel lines never intersect, so the system has no solution.

Method 2: SUBSTITUTION Because Equation 2 can be revised to $y = 3x + 2$, you can substitute $3x + 2$ for y in Equation 1.

$3x - y = 1$ Write Equation 1.

$3x - (3x + 2) = 1$ Substitute $3x + 2$ for y.

$-2 = 1$ Simplify. False statement.

The variables are eliminated and you have a statement that is not true regardless of the values of x and y. The system has no solution.

Exercises for Example 1

Choose a method to solve the linear system and tell how many solutions the system has.

1. $2x - y = 1$
$6x - 3y = 12$

2. $x + y = 5$
$3x + 3y = 7$

3. $2x + 6y = 6$
$x + 3y = -3$

EXAMPLE 2 *A Linear System with Many Solutions*

Use linear combinations to show that the linear system has infinitely many solutions.

$3x + y = 4$ Equation 1

$6x + 2y = 8$ Equation 2

NAME ______________________ DATE ____________

Reteaching with Practice

For use with pages 426–431

EXAMPLE 2 **SOLUTION**

You can multiply Equation 1 by -2.

$-6x - 2y = -8$	Multiply Equation 1 by -2.
$6x + 2y = 8$	Write Equation 2.
$0 = 0$	Add the equations.

The variables are eliminated and you have a statement that is true regardless of the values of x and y. The system has infinitely many solutions.

Exercises for Example 2

Choose a method to solve the linear system and tell how many solutions the system has.

4. $2x + 3y = 6$
$6x + 9y = 18$

5. $4x + 6y = 12$
$6x + 9y = 18$

6. $4x - 2y = 6$
$2x - y = 3$

EXAMPLE 3 ## *Modeling a Real-Life Problem*

An artist is buying art supplies. She buys 4 sketchpads and 2 palettes. She pays $16 for the supplies. The following week, at the same prices, she buys 2 sketchpads and one palette and pays $8. Can you find the price of one sketchpad? Explain.

SOLUTION

EXAMPLE 3

Let x represent the price of a sketchpad and let y represent the price of a palette. Determine the number of solutions of the linear system:

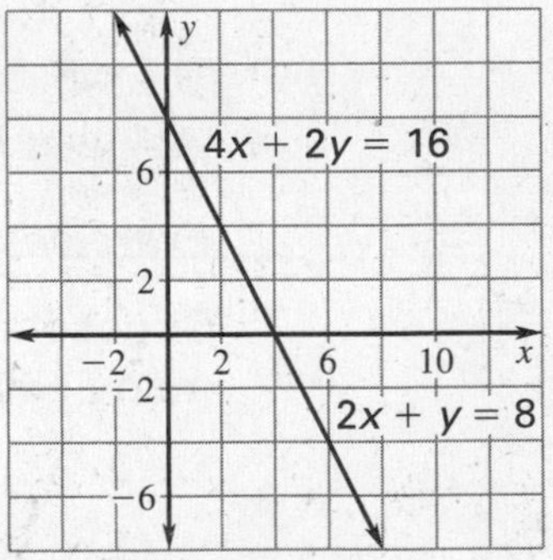

$4x + 2y = 16$	Equation 1
$2x + y = 8$	Equation 2

4x + 2y = 16
2x + y = 8
y
x
6
2
−2
−6
−2
2
6
10

Use the graphing method to identify the number of solutions for the linear system. Rewrite each equation in slope-intercept form and graph the linear system.

$y = -2x + 8$	Revised Equation 1
$y = -2x + 8$	Revised Equation 2

The equations represent the same line. Any point on the line is a solution. You cannot find the price of one sketchpad.

Exercise for Example 3

7. Rework Example 3, if the cost of the second purchase was $5 for one sketchpad and one palette.

LESSON 7.5

NAME ______________________________ DATE __________

Quick Catch-Up for Absent Students

For use with pages 425–431

The items checked below were covered in class on (date missed) ___________

Activity 7.5: Investigating Special Types of Linear Systems (p. 425)

____ **Goal:** Identify the number of solutions of a linear system by graphing.

____ Student Help: Look Back

Lesson 7.5: Special Types of Linear Systems

____ **Goal 1:** Identify linear systems as having one solution, no solution, or infinitely many solutions. (pp. 426–427)

Material Covered:

____ Student Help: Look Back

____ Example 1: A Linear System with No Solution

____ Student Help: Study Tip

____ Example 2: A Linear System with Many Solutions

____ Example 3: Identifying the Number of Solutions

____ **Goal 2:** Model real-life problems using a linear system. (p. 428)

Material Covered:

____ Example 4: Error Analysis

____ Other (specify) ______________________________

____

Homework and Additional Learning Support

____ Textbook (specify) pp. 429–431

____

____ *Reteaching with Practice* worksheet (specify exercises) ______________

____ *Personal Student Tutor* for Lesson 7.5

NAME ______________________________ DATE __________

Real-Life Application: When Will I Ever Use This?

For use with pages 426–431

Four Corners in Allegheny National Forest

Four northwestern Pennsylvania counties, Elk, McKean, Forest, and Warren, intersect in the Allegheny National Forest. The intersection point is known as the "Four Corners" and is marked by a stone monument. You can walk through four counties by simply walking around the monument.

Enclosing 513,161 acres, Allegheny National Forest is best known for its valuable hardwoods. Black cherry, yellow poplar, white ash, red maple, and sugar maple are included in this category. Black cherry harvested from Allegheny National Forest is of exceptional quality and is used by makers of fine furniture and veneers around the world.

Allegheny National Forest has abundant wildlife, providing opportunities to photograph or watch animals in their natural environment. Over 300 species of mammals inhabit forest lands including snowshoe hare, red and grey fox, beaver, white-tailed deer, and black bear. Visitors can also take advantage of the diverse water activities available at the 27-mile long Allegheny Reservoir.

In Exercises 1-3, use the following information.

You are going for a walk to Four Corners. On the way you hope to see some of the wildlife found in the Allegheny National Forest.

1. The path you take to Four Corners is represented by $8x - 4y = 68$. There is a flock of turkeys in front of you. They are walking in a line represented by the equation $3y + 12x = 27$. Is there a possibility you will cross paths with the flock of turkeys? Explain why or why not.

2. You continue on your trail and further ahead of you are two black bear cubs. They are following a path represented by the equation $2y - 4x = -34$. Is there a possibility you will cross paths with the cubs? Explain why or why not.

3. You are getting close to Four Corners. You take a short cut through the woods that is represented by the equation $5y - 3x = -21$. There is a small stream in the woods you are cutting through. The stream is represented by the equation $-6x + 10y = 50$. Will you have to cross the stream? Explain why or why not.

LESSON 7.5

NAME _______________ DATE _______________

Challenge: Skills and Applications

For use with pages 426–431

In Exercises 1–3, use the linear system.

$\frac{1}{2}x - \frac{3}{4}y = 5$

$kx - \frac{3}{5}y = 2$

1. For what values of k does the system have no solution?
2. For what values of k does the system have infinitely many solutions?
3. For what values of k does the system have exactly one solution?

In Exercises 4–5, suppose *a, b,* and *k* are nonzero numbers. Suppose you solve the system by linear combinations.

$ax + by = 5$

$kax + kby = 10$

4. Does the number of solutions the system has depend on the values of a and b? Does it depend on the value of k?
5. Describe the number of solutions in each possible case.

In Exercises 6–7, suppose you solve the system by multiplying the first equation by *d* and the second equation by *b* and then subtracting.

$ax + by = p$

$cx + dy = q$

6. What is the coefficient of x in the resulting equation?
7. State a relationship among the numbers a, b, c, and d that guarantees that the system does *not* have exactly one solution.

LESSON 7.6

TEACHER'S NAME ____________ CLASS ______ ROOM ______ DATE ______

Lesson Plan

2-day lesson (See *Pacing the Chapter,* TE pages 394C–394D) For use with pages 432–438

1. **Solve a system of linear inequalities by graphing.**
2. **Use a system of linear inequalities to model a real-life situation.**

State/Local Objectives ____________

✓ **Check the items you wish to use for this lesson.**

STARTING OPTIONS

____ Homework Check: TE page 429; Answer Transparencies
____ Warm-Up or Daily Homework Quiz: TE pages 432 and 431, CRB page 80, or Transparencies

TEACHING OPTIONS

____ Lesson Opener (Activity): CRB page 81 or Transparencies
____ Graphing Calculator Activity with Keystrokes: CRB pages 82–84
____ Examples: Day 1: 1–3, SE pages 432–433; Day 2: 4, SE page 434
____ Extra Examples: Day 1: TE page 433 or Transp.; Day 2: TE page 434 or Transp.; Internet
____ Closure Question: TE page 434
____ Guided Practice: SE page 435; Day 1: Exs. 1–8; Day 2: none

APPLY/HOMEWORK

Homework Assignment

____ Basic Day 1: 9–14, 16–26 even; Day 2: 30–36, 42, 45–52, 57, 58; Quiz 2: 1–14
____ Average Day 1: 9–14, 16–26 even; Day 2: 30–39, 42, 45–52, 57, 58; Quiz 2: 1–14
____ Advanced Day 1: 9–14, 16–26 even; Day 2: 30–39, 42–52, 57, 58; Quiz 2: 1–14

Reteaching the Lesson

____ Practice Masters: CRB pages 85–87 (Level A, Level B, Level C)
____ Reteaching with Practice: CRB pages 88–89 or Practice Workbook with Examples
____ Personal Student Tutor

Extending the Lesson

____ Applications (Interdisciplinary): CRB page 91
____ Challenge: SE page 437; CRB page 92 or Internet

ASSESSMENT OPTIONS

____ Checkpoint Exercises: Day 1: TE page 433 or Transp.; Day 2: TE page 434 or Transp.
____ Daily Homework Quiz (7.6): TE page 437, Transparencies
____ Standardized Test Practice: SE page 437; TE page 437; STP Workbook; Transparencies
____ Quizzes (7.4–7.6): SE page 438; CRB page 93

Notes ____________

LESSON 7.6

TEACHER'S NAME ____________________ CLASS ______ ROOM ______ DATE ______

Lesson Plan for Block Scheduling

1-day lesson (See *Pacing the Chapter,* TE pages 394C–394D) For use with pages 432–438

GOALS
1. **Solve a system of linear inequalities by graphing.**
2. **Use a system of linear inequalities to model a real-life situation.**

State/Local Objectives ____________________

CHAPTER PACING GUIDE	
Day	**Lesson**
1	7.1 (all); 7.2 (all)
2	7.3 (all)
3	7.4 (all)
4	7.5 (all); **7.6 (begin)**
5	**7.6 (end)**; Review Ch. 7
6	Assess Ch. 7; 8.1 (begin)

✓ Check the items you wish to use for this lesson.

STARTING OPTIONS
____ Homework Check: TE page 429; Answer Transparencies
____ Warm-Up or Daily Homework Quiz: TE pages 432 and 431, CRB page 80, or Transparencies

TEACHING OPTIONS
____ Lesson Opener (Activity): CRB page 81 or Transparencies
____ Graphing Calculator Activity with Keystrokes: CRB pages 82–84
____ Examples: Day 4: 1–3, SE pages 432–433; Day 5: 4, SE page 434
____ Extra Examples: Day 4: TE page 433 or Transp.; Day 5: TE page 434 or Transp.; Internet
____ Closure Question: TE page 434
____ Guided Practice: SE page 435; Day 4: Exs. 1–8; Day 5: none

APPLY/HOMEWORK
Homework Assignment (See also the assignment for Lesson 7.5.)
____ Block Schedule: Day 4: 9–14, 16–26 even; Day 5: 30–39, 42, 45–52, 57, 58; Quiz 2: 1–14

Reteaching the Lesson
____ Practice Masters: CRB pages 85–87 (Level A, Level B, Level C)
____ Reteaching with Practice: CRB pages 88–89 or Practice Workbook with Examples
____ Personal Student Tutor

Extending the Lesson
____ Applications (Interdisciplinary): CRB page 91
____ Challenge: SE page 437; CRB page 92 or Internet

ASSESSMENT OPTIONS
____ Checkpoint Exercises: Day 4: TE page 433 or Transp.; Day 5: TE page 434 or Transp.
____ Daily Homework Quiz (7.6): TE page 437, or Transparencies
____ Standardized Test Practice: SE page 437; TE page 437; STP Workbook; Transparencies
____ Quizzes (7.4–7.6): SE page 438; CRB page 93

Notes ____________________

LESSON **7.6**

NAME ____________________ DATE __________

Available as a transparency

WARM-UP EXERCISES

For use before Lesson 7.6, pages 432–438

In Exercises 1 and 2, tell whether the graph of the inequality will have a solid or a dashed boundary line.

1. $y \leq 2x + 3$

2. $-2x + 1 > y$

3. Which inequality's graph is included in the graph of one of the other inequalities?

$2x + 3y \leq 8$ $\quad$ $3x + 2y \leq 7$

$4x + 6y \leq 9$ $\quad$ $x + y > 6$

DAILY HOMEWORK QUIZ

For use after Lesson 7.5, pages 425–431

The diagram shows the graphs of the following three equations.

A: x − 2y = 2

B: −2x + 4y = 8

C: 3x + 2y = 2

How many solutions are there for the system formed by the pair of equations?

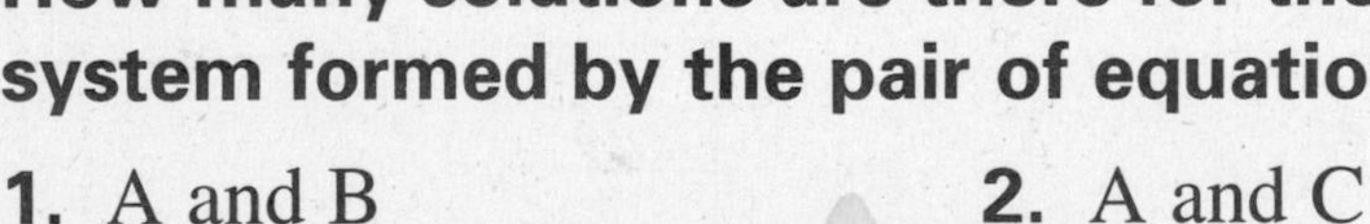

1. A and B

2. A and C

Solve the linear system. Tell how many solutions the system has.

3. $3x + 4y = 5$
$-2x + y = 4$

4. $7x - 3y = 4$
$-14x + 6 = -8$

5. For 4 hamburgers and 4 sodas you pay \$8. For 6 hamburgers and 6 sodas you pay \$12. Can you use this information to find the price of 1 hamburger and the price of 1 soda? Explain.

LESSON

7.6

NAME ______________________ DATE __________

Available as a transparency

Activity Lesson Opener

For use with pages 432–438

SET UP: Work with a partner.

Two lines are shown on the graph at the right. The equations of these lines are $y = 2x + 1$ and $y = -x + 2$.

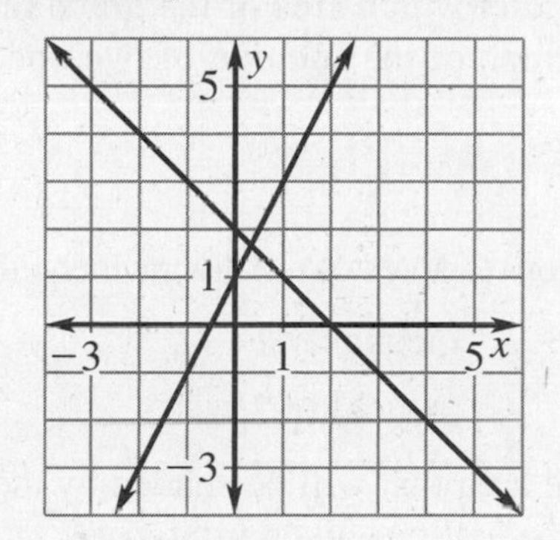

1. Graph the point $(-2, 3)$ on the same grid.
2. Shade the region bounded by the two lines that contains this point.
3. Write the linear inequality for the half-plane that contains the point $(-2, 3)$ and is bounded by the line $y = 2x + 1$.
4. Repeat Question 3 using the line $y = -x + 2$.
5. What is true about the shaded region and the linear inequalities you wrote in Questions 3 and 4?

Two lines are shown on the graph at the right. The equations of these lines are $y = 3x - 2$ and $y = \frac{1}{2}x + 2$.

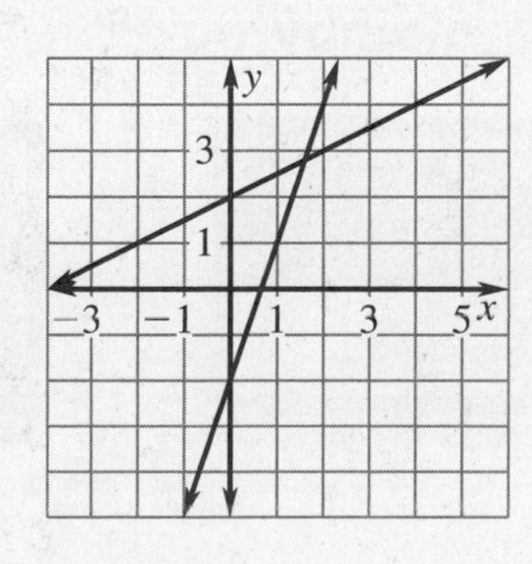

6. Graph the point $(1, -4)$ on the same grid.
7. Shade the region bounded by the two lines that contains this point.
8. Write the linear inequality for the half-plane that contains the point $(1, -4)$ and is bounded by the line $y = 3x - 2$.
9. Repeat Question 8 using the line $y = \frac{1}{2}x + 2$.
10. What is true about the shaded region and the linear inequalities you wrote in Questions 8 and 9?

LESSON 7.6

NAME ______________________ DATE ________

Graphing Calculator Activity

For use with pages 432–438

GOAL **To predict which area of a graph represents the solution of a system of linear inequalities**

By observing the linear inequalities in a system of linear inequalities, it is possible to predict which area of the graph represents the solution of the system. The graph of the solution of two linear inequalities is shown at the left.

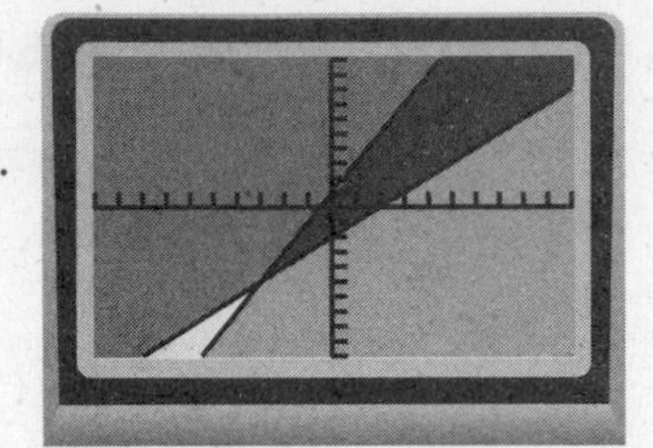

Activity

1. Listed below is a system of linear inequalities.

   $y \geq x + 3$ Inequality 1

   $y \geq 3$ Inequality 2

2. Predict the area that will be shaded by the graph representing the system of linear inequalities given in Step 1.
3. Use your graphing calculator to enter the two inequalities.
4. Graph the two inequalities using the shading feature of your calculator.
5. Does your graph verify your prediction?

Exercises

1. Predict the area of the graph that will be shaded by the system of linear inequalities. Use your graphing calculator to check your prediction.

a. $y \leq -3$
$y \leq x - 5$

b. $y \geq x + 5$
$y \leq x + 7$

c. $y \geq -2x - 3$
$y \geq 3x - 3$

2. Using the information in the graph, complete the system of inequalities with $\leq$ or $\geq$.

a. y_1 __?__ -2
y_2 __?__ $-3x + 3$

b. y_1 __?__ $-x - 4$
y_2 __?__ $-5x - 9$

c. y_1 __?__ $4x - 7$
y_2 __?__ $-6x + 5$

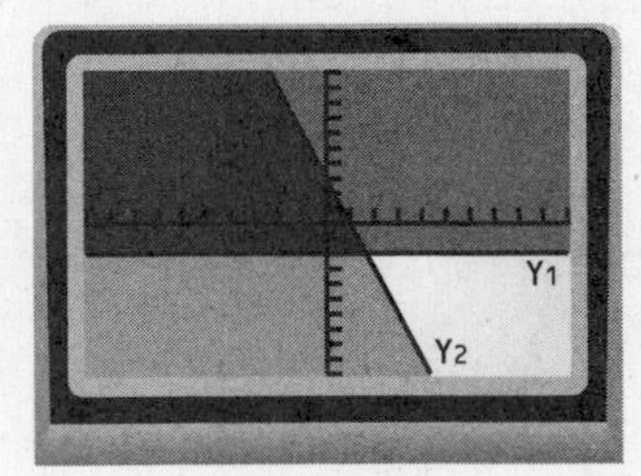

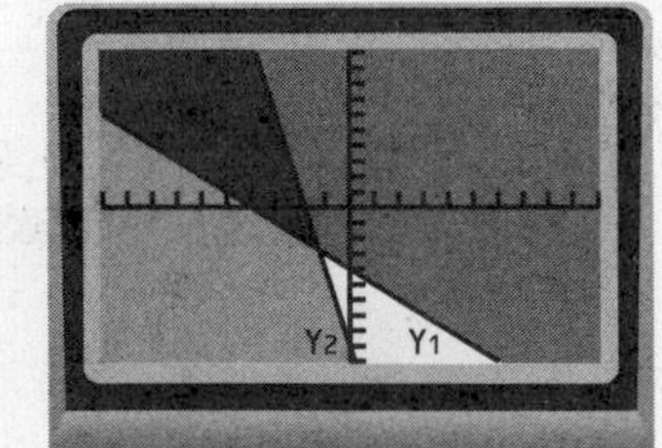

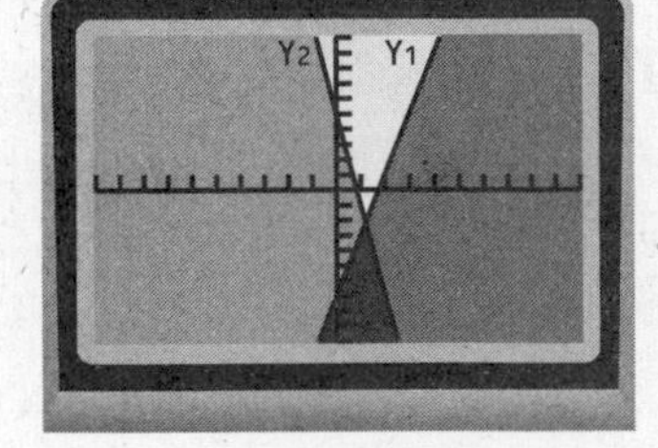

NAME ______________________________ DATE ____________

Graphing Calculator Activity

For use with pages 432–438

TI-82

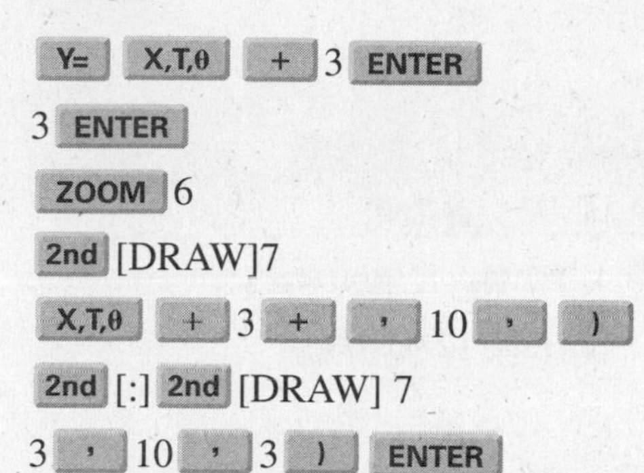

Y= X,T,θ + 3 ENTER

3 ENTER

ZOOM 6

2nd [DRAW]7

X,T,θ + 3 + , 10 ,)

2nd [:] 2nd [DRAW] 7

3 , 10 , 3) ENTER

TI-83

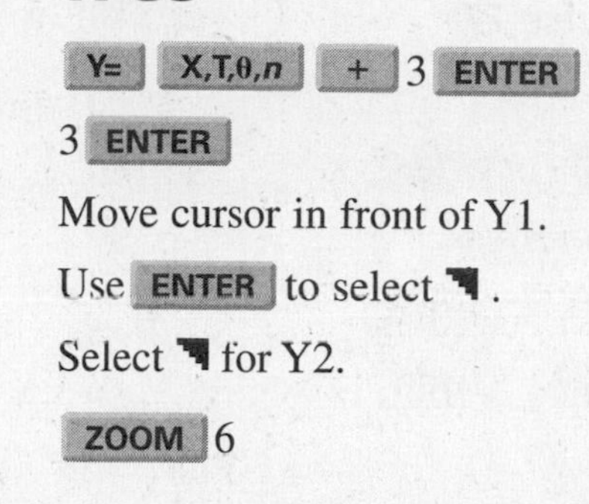

Y= X,T,θ,n + 3 ENTER

3 ENTER

Move cursor in front of Y1.

Use ENTER to select ◥.

Select ◥ for Y2.

ZOOM 6

SHARP EL-9600c

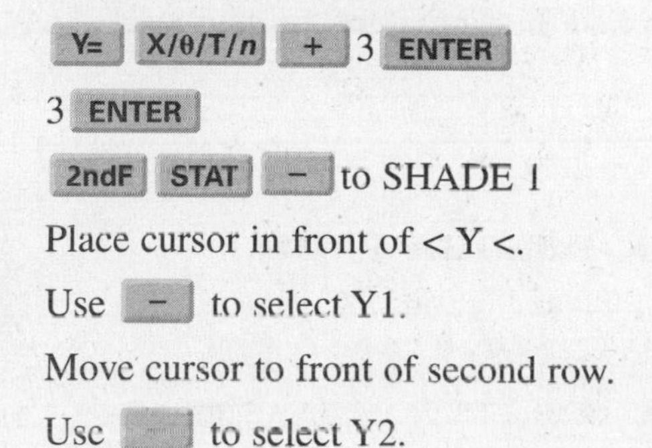

Y= X/θ/T/n + 3 ENTER

3 ENTER

2ndF STAT − to SHADE 1

Place cursor in front of < Y <.

Use − to select Y1.

Move cursor to front of second row.

Use − to select Y2.

CASIO CFX-9850GA PLUS

From the main menu, choose GRAPH.

Enter x-values in List 1.

F3 F6 F3 X,θ,T + 3 EXE

F3 F6 F3 3 EXE

SHIFT F3 F3

EXIT F6

NAME ______________________ DATE ____________

Graphing Calculator Activity Keystrokes

For use with pages 436

Keystrokes for Exercise 27

TI-82

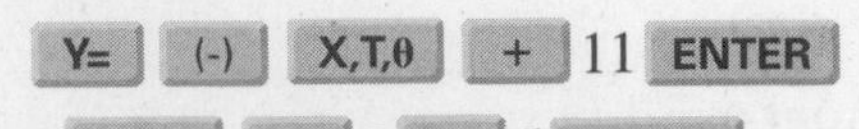

5 [X,T,θ] [÷] 3 [−] 5 [ENTER]

[WINDOW] [ENTER] [(-)] 1 [ENTER] 12

[ENTER] 1 [ENTER] [(-)] 8 [ENTER] 12

[ENTER] 1 [ENTER]

[2nd] [DRAW]7 [2nd] [Y-VARS] 1 2 [,] [2nd]

[Y-VARS] 1 1 [,] 1 [,] 0 [)] [ENTER]

[TRACE]

Use the cursor keys, [◀] and [▶], to move the trace cursor to find the vertices of the solution region.

TI-83

[Y=] [(-)] [X,T,θ,n] [+] 11 [ENTER]

5 [X,T,θ,n] [÷] 3 [−] 5 [ENTER]

[WINDOW] [(-)] 1 [ENTER] 12 [ENTER] 1

[ENTER] [(-)] 8 [ENTER] 12 [ENTER] 1

[ENTER]

[2nd] [DRAW]7 [VARS] [▶] 1 2 [,] [VARS] [▶]

1 1 [,] 0 [)] [ENTER]

[TRACE]

Use the cursor keys, [◀] and [▶], to move the trace cursor to find the vertices of the solution region

SHARP EL-9600c

[Y=] [(-)] [X/θ/T/n] [+] 11 [ENTER]

5 [X/θ/T/n] [÷] 3 [−] 5 [ENTER]

[WINDOW] [(-)] 1 [ENTER] 12 [ENTER] 1

[ENTER] [(-)] 8 [ENTER] 12 [ENTER] 1

[ENTER]

[2ndF] [DRAW][A]7 [VARS] [A] [ENTER]

[A] 2 [,] [VARS] [ENTER] 1 [,] 0 [)] [ENTER]

[TRACE]

Use the cursor keys, [◀] and [▶], to move the trace cursor to find the vertices of the solution region.

CASIO CFX-9850GA PLUS

From the main menu, choose GRAPH.

[F3] [F6] [F4] [(-)] [X,θ,T] [+] 11 [EXE]

[F3] [F6] [F3] 5 [X,θ,T] [÷] 3 [−] 5 [EXE]

[F3] [F4] 0 [EXE]

[SHIFT] [F3] 0 [EXE] 12 [EXE] 1 [EXE] [(-)]

8 [EXE] 12 [EXE] 1 [EXE] [EXIT]

[F6] [F1]

Use the cursor keys, [◀] and [▶], to move the trace cursor to find the vertices of the solution region.

LESSON

7.6

NAME ________________________________ DATE __________

Practice A

For use with pages 432–438

Match the system of linear inequalities with its graph.

A. $x + y \leq 4$
$x + y \geq -2$

B. $x + 2y \leq 4$
$-2x + y \geq -2$

C. $x + y \geq -2$
$-2x + y \geq -2$

1.

2.

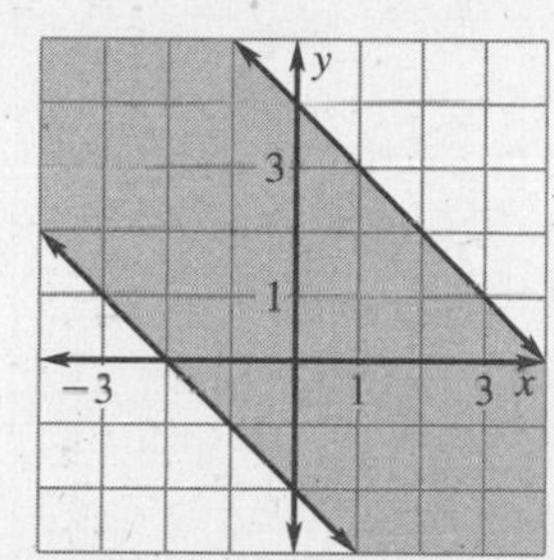

3.

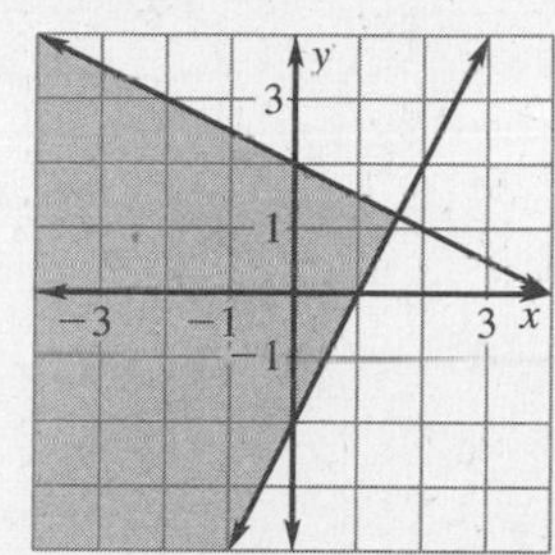

Write a system of linear inequalities that defines the shaded region.

4.

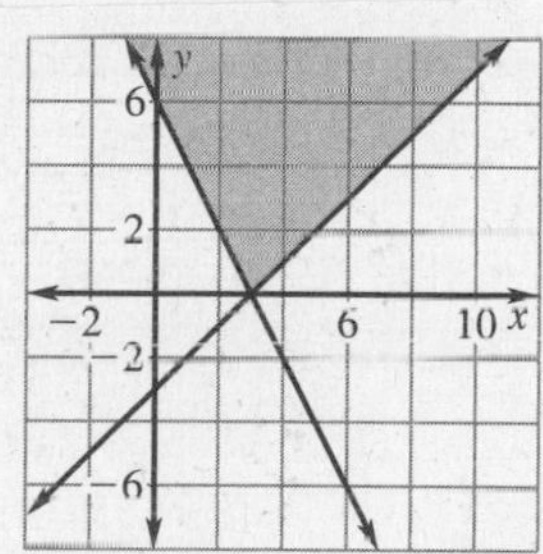

5.

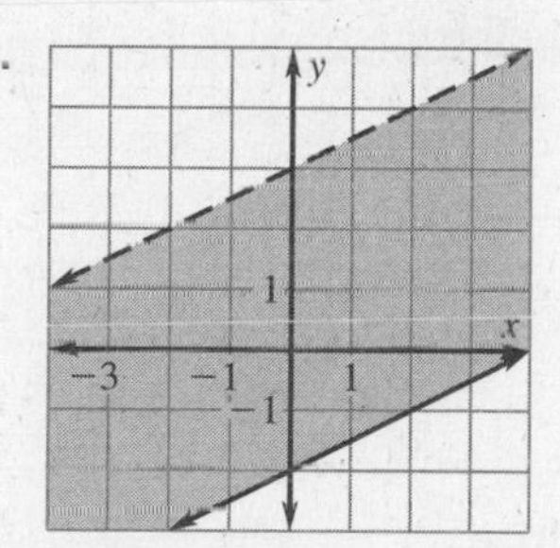

6.

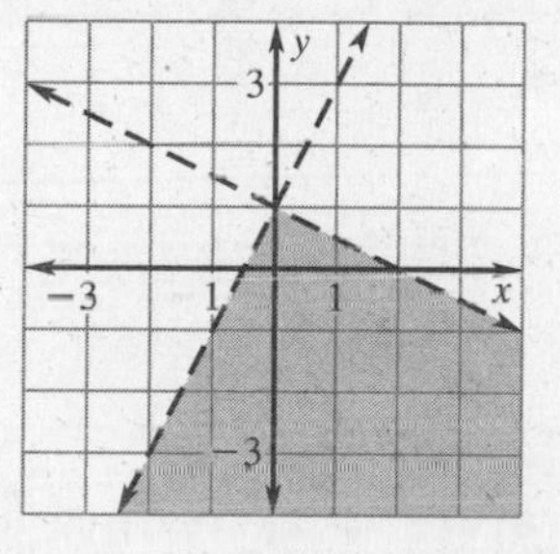

Graph the system of linear inequalities.

7. $y \geq 2$
$x < -3$

8. $y < 2x + 1$
$y \geq \frac{1}{2}x$

9. $x + y \geq 4$
$-3x + y < 1$

10. $2x + 3y < 4$
$2x + 3y > -9$

11. $3x - 4y > 2$
$3x - y \geq 2$

12. $x \geq 0$
$x \geq 0$
$y > x - 2$

13. *Study Time* You need at least 3 hours to do your English and history homework. It is 12:00 P.M. on Sunday and your friend wants you to go to the movies at 7:00 P.M. Write a system of linear inequalities that shows the number of hours you could spend doing homework for each subject if you go to the movies. Graph your result.

14. *Ordering Cups* You work at a frozen yogurt shop during the summer. You need to order 5-ounce and 8-ounce cups. The storage room will only hold 10 more boxes. A box of 5-ounce cups costs $100 and a box of 8-ounce cups costs $150. A maximum of $1200 is budgeted for yogurt cups. Write a system of linear inequalities that shows the number of boxes of 5-ounce and 8-ounce cups that could be bought. Graph your result.

15. *Geometry Connection* Write a system of linear inequalities that defines the polygonal region shown.

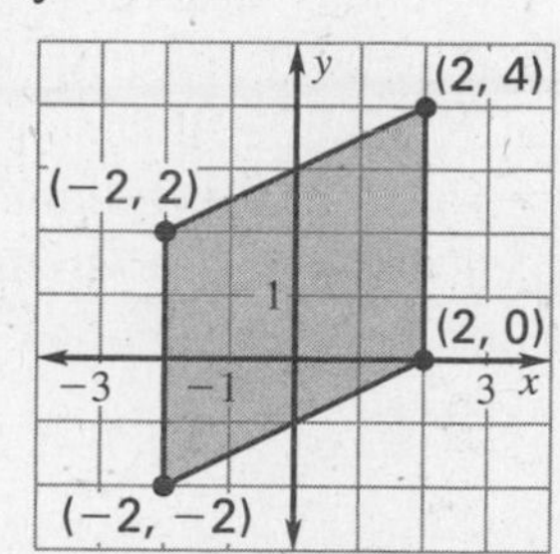

NAME ______________________________ DATE __________

Practice B

For use with pages 432–438

Match the system of linear inequalities with its graph.

A. $x + 2y \le 2$
$x + 2y \ge -2$

B. $x - 2y \ge 4$
$2x - y \ge -2$

C. $2x + 2y \ge -4$
$-2x - y \ge 2$

1.

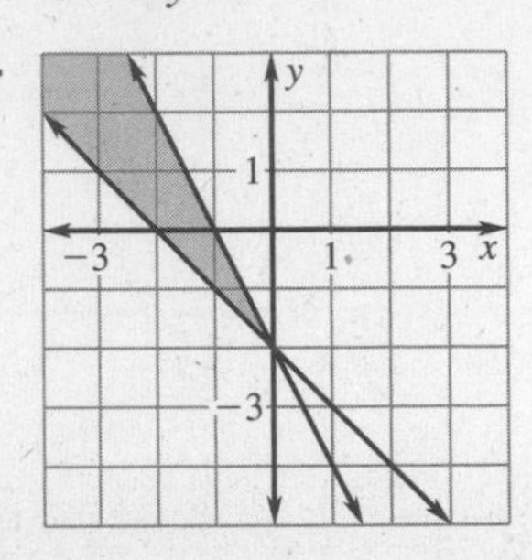

2.

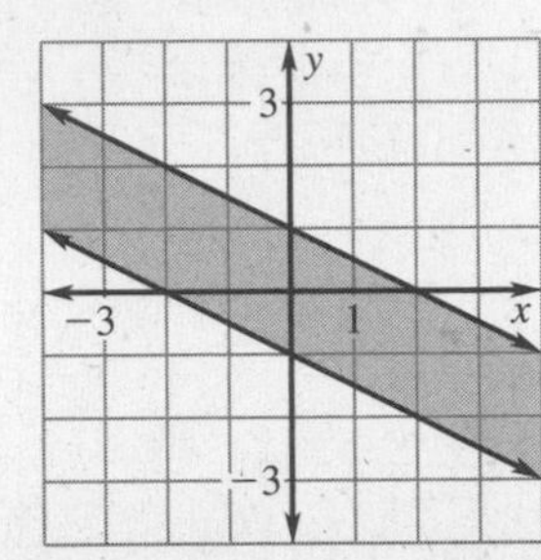

3.

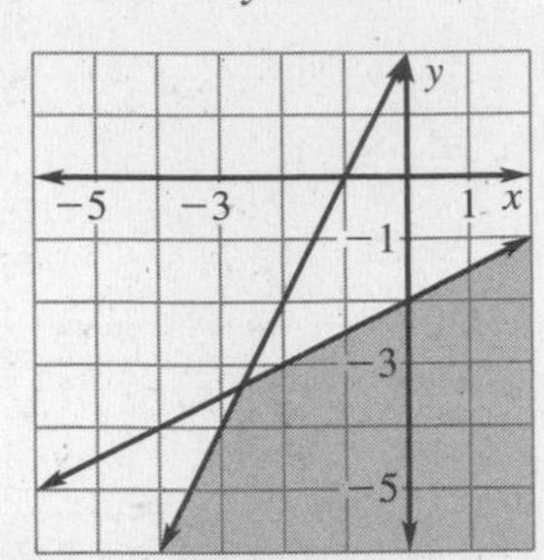

Write a system of linear inequalities that defines the shaded region.

4.

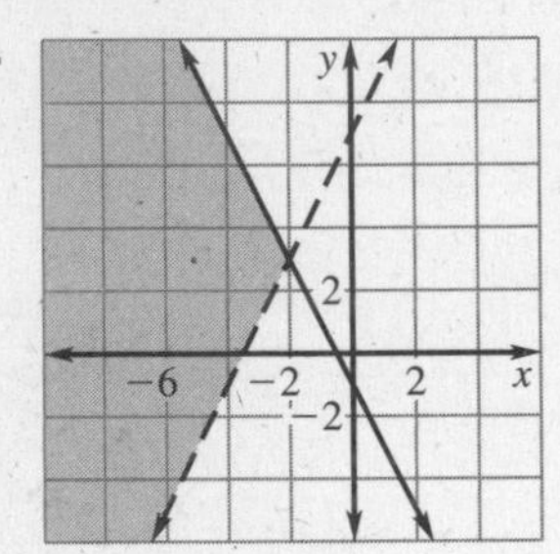

5.

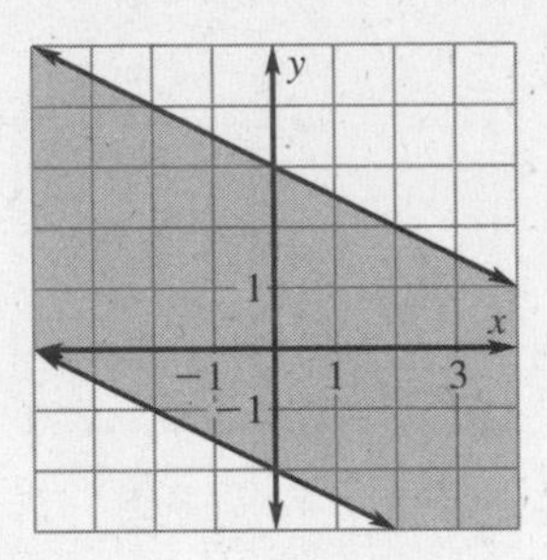

6.

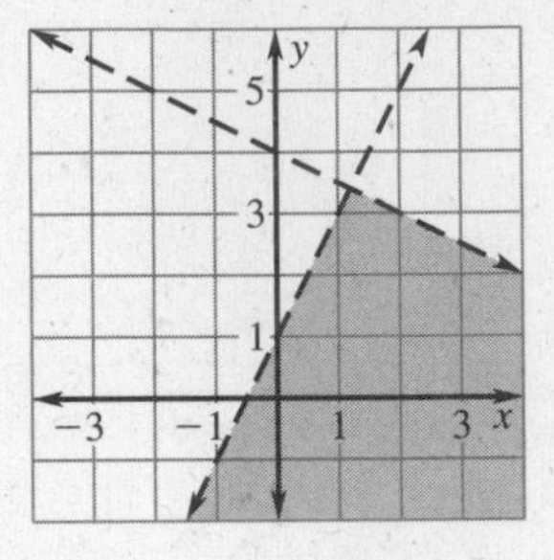

Graph the system of linear inequalities.

7. $4x + 2y \ge -6$
$8x + y < 3$

8. $2x + 3y < 1$
$2x + 3y > -9$

9. $3x - 6y > 2$
$3x - y \ge 2$

10. $x \le 0$
$y \ge 0$
$y \le 5$
$x > -6$

11. $2x + y \le 4$
$-3x + y < 3$
$y \ge -4$

12. $x + y < 3$
$-x - 3y \le 2$
$2x + \frac{1}{4}y > -1$

Plot the points and draw line segments connecting the points to create the polygon. Then write a system of linear inequalities that defines the polygonal region.

13. Rectangle: $(-1, 5), (-1, -1), (3, -1), (3, 5)$

14. Triangle: $(-2, 4), (4, 1), (-2, -1)$

15. *Study Time* You need at least 3 hours to do your English and history homework. It is 12:30 P.M. on Sunday and your friend wants you to go to the movies at 7:00 P.M. Write a system of linear inequalities that shows the number of hours you could spend doing homework for each subject if you go to the movies. Graph your result.

16. *Ordering Cups* You work at a frozen yogurt shop during the summer. You need to order 5-ounce and 8-ounce cups. The storage room will only hold 10 more boxes. A box of 5-ounce cups costs \$100 and a box of 8-ounce cups costs \$150. A maximum of \$1250 is budgeted for yogurt cups. Write a system of linear inequalities that shows the number of boxes of 5-ounce and 8-ounce cups that could be bought. Graph your result.

LESSON 7.6

NAME ____ DATE ____________

Practice C

For use with pages 432–438

Write a system of linear inequalities that defines the shaded region.

1. 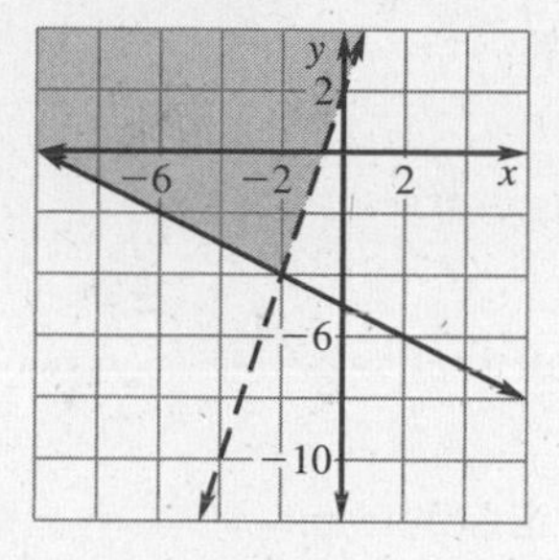

2. 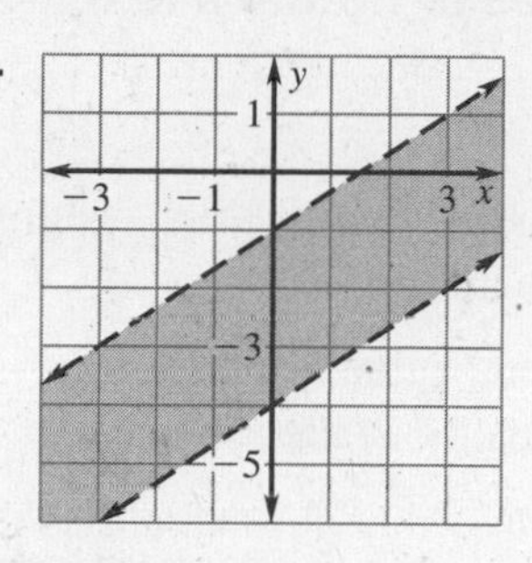

3. 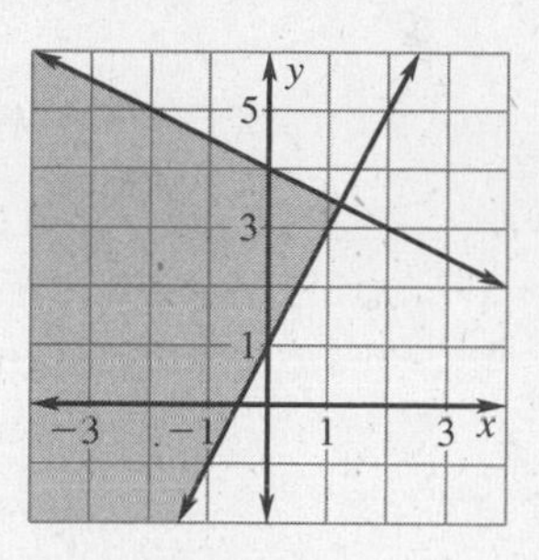

Graph the system of linear inequalities.

4. $4x + 2y \geq -6$
   $8x + y < 0$

5. $2x + 3y < 4$
   $2x + 3y > -5$

6. $3x - 5y > 2$
   $3x - y \geq 2$

7. $x \geq 0$
   $y \leq 0$
   $y < \frac{2}{5}x - 2$

8. $x \geq 0$
   $y \leq 0$
   $y > -7$
   $x \leq 6$

9. $4x - y \geq 2$
   $x \leq 4$
   $y \leq -2$

10. $-2x + 5y < 5$
    $x + 2y < 3$
    $x < 4$

11. $-3x + 4y < -7$
    $\frac{3}{4}x - y \leq 5$
    $y \leq 0$
    $x \geq 0$

12. $5x - y > -4$
    $5x + y \geq -6$
    $5x - 5y < 10$
    $5x - \frac{1}{3}y < 2$

Find the coordinates of the vertices of the solution region.

13. $x + y \geq 3$
    $-2x + y \leq 3$
    $3x + y \leq 13$

14. $x + 4y \leq 17$
    $4x - 5y \leq 5$
    $y \leq 4$
    $x \geq 0$

15. $2x + y \leq 6$
    $2x + 3y \geq 0$
    $y \leq 4$
    $x \leq 3$

Plot the points and draw the line segments connecting the points to create the polygon. Then write a system of linear inequalities that defines the polygonal region.

16. Rectangle: $(2, 1), (-1, 4), (-5, 0), (-2, -3)$

17. Triangle: $(-2, 4), (4, 1), (-4, -1)$

18. ***Study Time*** You need at least 3 hours to do your English and history homework. You need to spend at least twice as much time on your history homework as your English homework. It is 12:00 noon on Sunday and your friend wants you to go to the movies at 7:00 P.M. Write a system of linear inequalities that shows the number of hours you could spend doing homework for each subject if you go to the movies. Graph your result.

19. ***Ordering Cups*** You work at a frozen yogurt shop during the summer. You need to order 5-ounce and 8-ounce cups. The storage room will only hold 10 more boxes. A box of 5-ounce cups costs \$125 and a box of 8-ounce cups costs \$150. A maximum of \$1300 is budgeted for yogurt cups. Write a system of linear inequalities that shows the number of boxes of 5-ounce and 8-ounce cups that could be bought. Graph your result.

LESSON **7.6**

NAME ______________________________ DATE __________

Reteaching with Practice

For use with pages 432–438

GOAL **Solve a system of linear inequalities by graphing and use a system of linear inequalities to model a real-life situation**

VOCABULARY

Two or more linear inequalities form a **system of linear inequalities** or simply a **system of inequalities.**

A **solution** of a system of linear inequalities is an ordered pair that is a solution of each inequality in the system.

The **graph** of a system of linear inequalities is the graph of all solutions of the system.

Lesson 7.6

EXAMPLE 1 *A Triangular or Quadrilateral Solution Region*

Graph the system of linear inequalities.

$x - y \geq 0$	Inequality 1
$x + y \geq 0$	Inequality 2
$x \leq 3$	Inequality 3
$y \leq 2$	Inequality 4

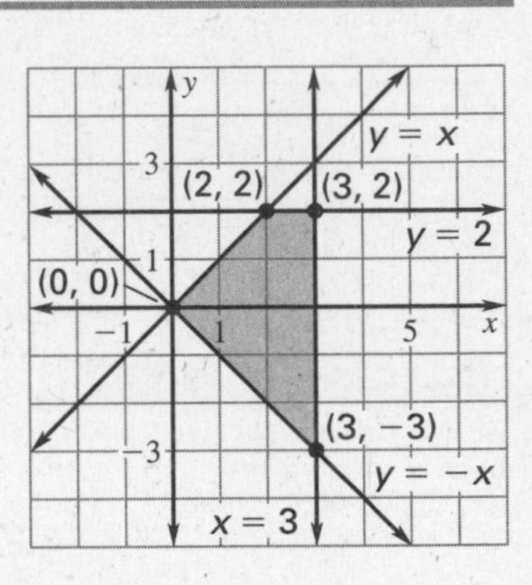

SOLUTION

Graph all four inequalities in the same coordinate system. The graph of the system is the overlap, or intersection, of the four half-planes shown.

When graphing a system of linear inequalities, find each corner point (or vertex). The graph of the system for Example 1 has four corner points: (0, 0), (2, 2), (3, 2), and (3, −3).

Exercises for Example 1

Graph the system of linear inequalities.

1. $x + y \leq 5$
$x > 1$
$y > -1$

2. $2x + 3y < 6$
$2x + y \leq 2$

3. $y \geq x - 1$
$y \leq -x + 1$
$y \geq -1$
$x \geq -1$

LESSON 7.6 CONTINUED

NAME ______________________ DATE __________

Reteaching with Practice

For use with pages 432–438

EXAMPLE 2 *Writing a System of Linear Inequalities*

Suppose that you can spend no more than \$72 for compact discs and videos. Discs cost \$18 each and videos cost \$9 each. Write a system of linear inequalities that shows the various numbers of discs and videos that you can buy.

SOLUTION

Verbal Model

[Number of discs] ≥ 0

[Number of videos] ≥ 0

[Number of discs] · [Cost of a disc] + [Number of videos] · [Cost of a video] ≤ 72

Labels

Number of discs $= x$	(no units)
Number of videos $= y$	(no units)
Cost of a disc $= 18$	(dollars)
Cost of a video $= 9$	(dollars)

Algebraic Model

$x \geq 0$	Inequality 1
$y \geq 0$	Inequality 2
$18x + 9y \leq 72$	Inequality 3

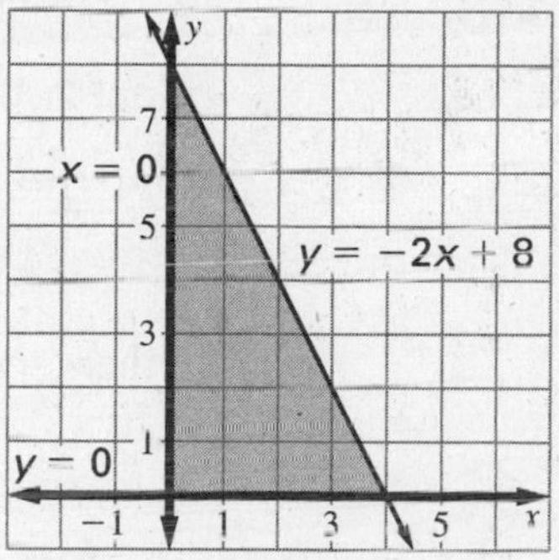

$x = 0$, $y = 0$, $y = -2x + 8$

The graph of the system of inequalities is shown. Any point in the shaded region is a solution of the system. Because you cannot buy a fraction of a disc or video, only ordered pairs of integers in the shaded region will answer the problem.

Exercises for Example 2

4. Rework Example 3 if you can spend no more than \$90.
5. Rework Example 3 if discs cost \$16 each and videos cost \$8 each.

Lesson 7.6

NAME ______________________________ DATE __________

Quick Catch-Up for Absent Students

For use with pages 432–438

The items checked below were covered in class on (date missed) __________

Lesson 7.6: Solving Systems of Linear Inequalities

____ **Goal 1:** Solve a system of linear inequalities by graphing. (pp. 432–433)

Material Covered:

____ Example 1: A Triangular Solution Region

____ Example 2: Solution Region Between Parallel Lines

____ Student Help: Study Tip

____ Example 3: A Quadrilateral Solution Region

Vocabulary:

system of linear inequalities, p. 432
solution of a system of linear inequalities, p. 432
graph of a system of linear inequalities, p. 432

____ **Goal 2:** Use a system of linear inequalities to model a real-life situation. (p. 434)

Material Covered:

____ Example 4: Writing a System of Linear Inequalities

____ Other (specify) ______________________________

____

Homework and Additional Learning Support

____ Textbook (specify) pp. 435–438

____

____ Internet: Extra Examples at www.mcdougallittell.com

____ *Reteaching with Practice* worksheet (specify exercises) __________

____ *Personal Student Tutor* for Lesson 7.6

NAME ______________________ DATE __________

Interdisciplinary Application

For use with pages 432–438

Salt Water Fish

BIOLOGY Fish are vertebrates that live in water. There are more kinds of fish than all other kinds of water and land vertebrates put together. The various kinds of fish differ so greatly in shape, color, and size that it is hard to believe they all belong to the same group of animals.

About 13,300 species–or about three-fifths of all known fish– live in the ocean (saltwater). Most of the saltwater fish are suited to a particular type of environment and cannot survive in one much different from that type. Water temperature is one of the chief factors in determining where a fish can live.

Many saltwater species live where the water is always warm. The warmest parts of the ocean are the shallow tropical waters around the coral reefs. More than a third of all known saltwater species live around the coral reefs in the Indian and Pacific Oceans. Coral reefs swarm with angelfish, butterfly fish, parrot fish, and thousands of other species with fantastic shapes and brilliant colors. Barracudas, groupers, and sharks prowl the clear coral waters in search of prey.

In Exercises 1-3, use the following information.

Your biology class wants to set up a salt water fish tank. The class definitely likes eels, but the teacher suggests no more than 4 eels in the tank. The tank will hold no more than 20 fish. The tank can support no more than 15 eels and fish together.

1. Write a system of linear inequalities to represent the capacity of the fish tank.
2. Graph your system of linear inequalities.
3. There are 12 fish in the tank. How many eels can be added to the tank?
4. In a second tank, the class decides to have two kinds of fish–parrot fish and lion fish. Your teacher suggests not more than 30 fish in the tank. Your class decides to have at least 5 lion fish and at least 15 parrot fish. Write and graph a system of linear inequalities to represent the number of fish in the tank.

NAME ______________________ DATE __________

Challenge: Skills and Applications

For use with pages 432–438

In Exercises 1–2, graph the system of inequalities.

1. $y < x + 4$
   $y < -\frac{3}{2}x + 4$
   $y \geq \frac{1}{2}|x|$

2. $|x| < 3$
   $y > x - 1$
   $y < x + 1$

In Exercises 3–5, use the following information.

Teresa Sanchez sells two sizes of outdoor doghouses: large and small. The large size requires 12 board-feet of lumber and takes 3 hours to build. The small size requires 8 board-feet of lumber and takes 1 hour to build. Teresa can use 48 board-feet of lumber each day and plans to spend at most 9 hours per day building dog houses.

3. Model the situation above. Your algebraic model should be a system of four inequalities. (Remember that Teresa cannot build a negative number of dog houses.)
4. Graph the system of inequalities from Exercise 3.
5. Teresa sells her large dog houses for \$70 each and her small ones for \$30 each. What numbers of each kind should she make per day in order to maximize her income from sales? (*Hint:* The maximum income must occur at one of the vertices of the graph.)

LESSON 7.6

NAME ______________________ DATE ________

Quiz 2

For use after Lessons 7.4–7.6

1. Solve the linear system by any method. *(Lesson 7.4)*

$2x + y = 6$

$2x + 3y = 10$

2. Use the graphing method to solve the linear system and tell how many solutions the system has. *(Lesson 7.5)*

$2x + 4y = 8$

$3x + 6y = 12$

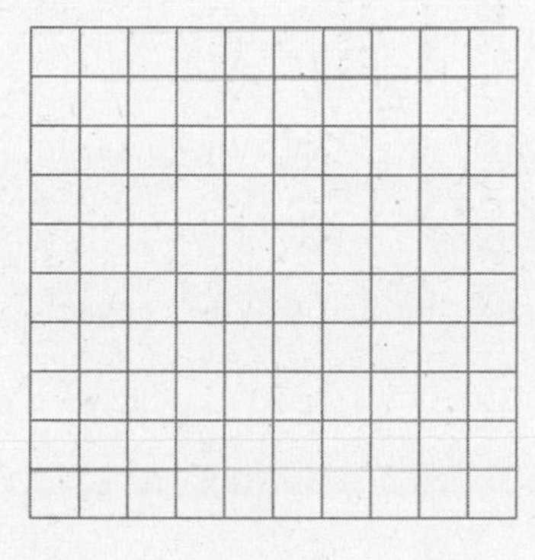

3. Use the substitution method or linear combinations to solve the linear system and tell how many solutions the system has. *(Lesson 7.5)*

$y = 6x + 5$

$6x - y = 7$

4. Graph the system of linear inequalities. *(Lesson 7.6)*

$x + y > 7$

$3x + y \leq 6$

5. Plot the points $(-6, 3)$, $(0, 3)$, $(-6, 0)$ and draw line segments connecting the points to create the polygon. Then write a system of linear inequalities that defines the polygonal region. *(Lesson 7.6)*

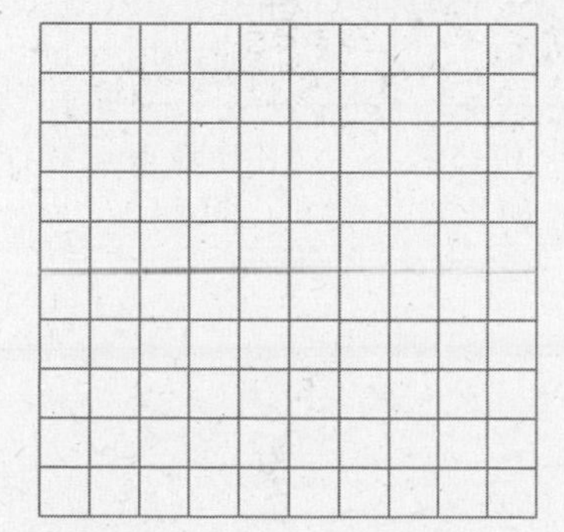

Answers

1. ____________
2. ____________
3. ____________
4. ____________
5. Use grid at left.

____

Lesson 7.6

NAME ______________________ DATE __________

Chapter Review Games and Activities

For use after Chapter 7

Solve the following systems of equations using the graphing, substitution, or linear combination method. Find the answer in the boxes at the bottom of the page. Cross out the box containing each correct answer. The remaining words will answer the question . . .

What catastrophe makes your kitten happy?

1. $8x + 3y = 3$
$4x + 5y = -23$

2. $x + y = 7$
$2x + 2y = 14$

3. $4x + 2y = 0$
$5x + 2y = -3$

4. $x + y = 6$
$x - 3y = 10$

5. $6x + 7y = 26$
$2x - y = 22$

6. $3x - 8y = -4$
$5x + 4y = 28$

7. $4x + 3y = -4$
$3x + 2y = -1$

8. $3x - 4y = 14$
$9x - 12y = 48$

9. $5x - 8y = 10$
$6x - 7y = -1$

10. $3x - 5y = 20$
$6x - 10y = 20$

11. $6x + 5y = -4$
$5x - 8y = 21$

12. $x - 3y = 6$
$4x - 12y = 24$

(4, 2) LET	(−6, −5) CHAIN	All real nos. THE	(−3, 6) HAVE
No solution CAT	(−5, −6) CRY	(3, −7) ONE	(1, −2) OUT
(2, −1) OVER	(5, −8) LITTER	(9, −4) DOG	(3, −6) SPILT
No solution NEW	(7, −1) OUTSIDE	(7, 1) MILK	All real nos. OLD

When you ______ ______ ______ ______.

NAME ______________________________ DATE ____________

Chapter Test A

For use after Chapter 7

Is the ordered pair a solution of the system of linear equations?

1. $x + y = 5$ $(0, 5)$
$-5x + 2y = 10$

2. $-x + y = -3$ $(4, 1)$
$x + 3y = 6$

Graph and check to solve the linear system.

3. $-x + y = 3$
$x + y = 5$

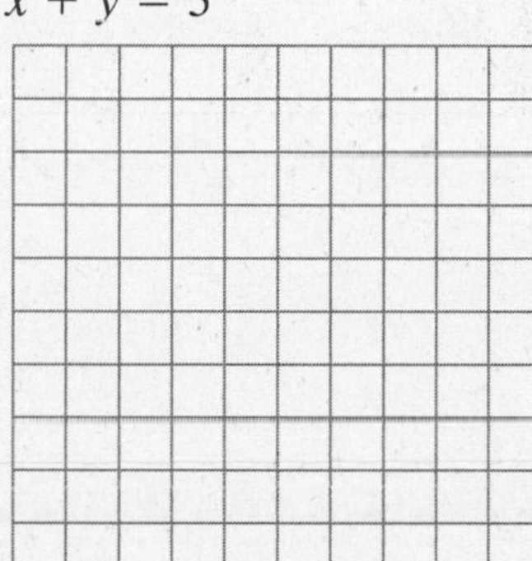

4. $x = 4$
$y = 2$

Use the substitution method to solve the linear system.

5. $y = x + 2$
$3x + 2y = 9$

6. $y - 3 = x$
$-2x + y = 6$

7. You are selling tickets for a high school basketball game. Student tickets cost \$3 and general admission tickets cost \$5. You sell 350 tickets and collect \$1450. How many of each type of ticket did you sell?

Use linear combinations to solve the linear system.

8. $x + y = 5$
$x - y = 3$

9. $x + y = 5$
$2x + y = 6$

10. A music store is selling compact discs for \$11.50 and \$7.50. You buy 12 discs and spend a total of \$110. How many compact discs that cost \$11.50 did you buy?

Solve the system using the method of your choice and tell how many solutions the system has.

11. $x + 2y = 5$
$2x - 2y = 4$

12. $x + y = 1$
$x + y = 3$

1. ____________
2. ____________
3. Use grid at left.
4. Use grid at left.
5. ____________
6. ____________
7. ____________
8. ____________
9. ____________
10. ____________
11. ____________
12. ____________

NAME ______________________________ DATE ____________

Chapter Test A

For use after Chapter 7

In Questions 13–16, match the system of linear inequalities with its graph.

A. $-x + y < 2$
$-x + y \geq -2$

B. $-x + y > 2$
$x + y \leq -2$

C. $-x + y > -2$
$x + y \leq 2$

D. $-x + y \leq 2$
$x + y < 2$

13.

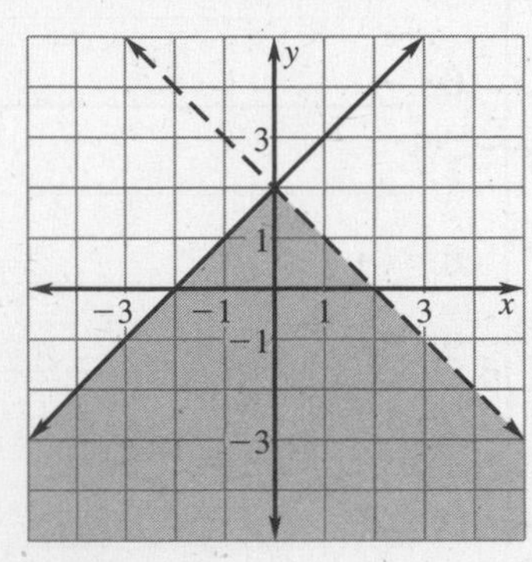

14.

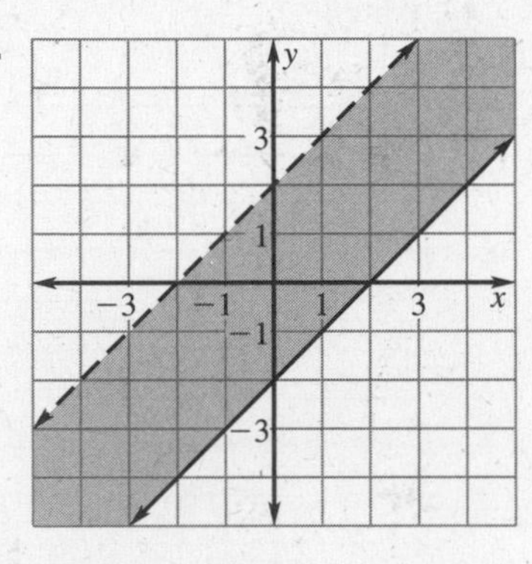

15.

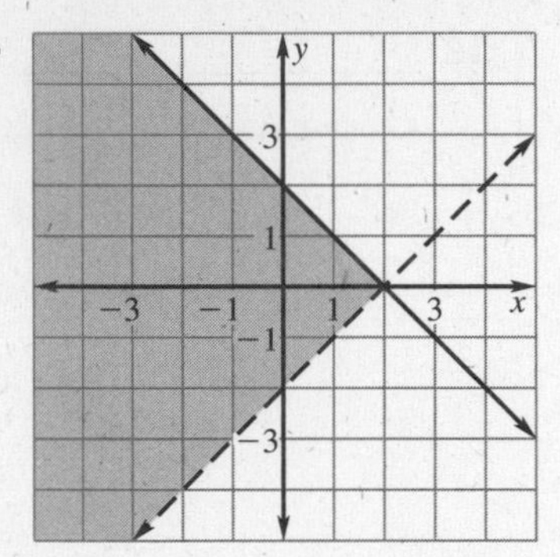

16.

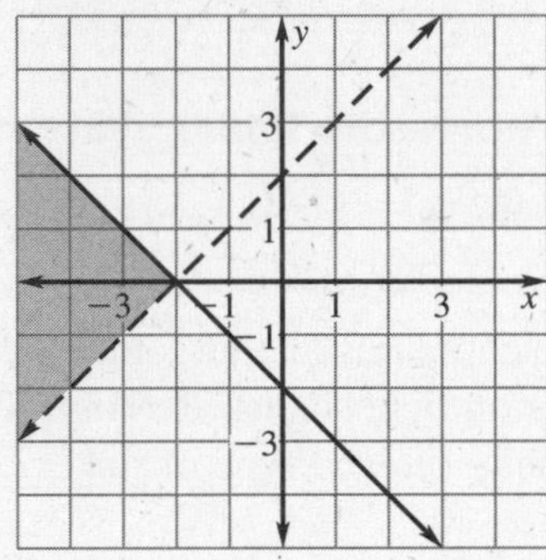

Graph the system of linear inequalities.

17. $y \geq -2$
$x < 2$

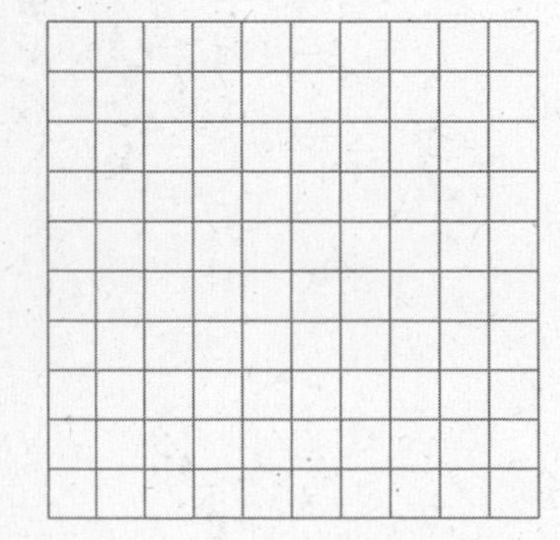

18. $y < x + 1$
$y \geq 3$

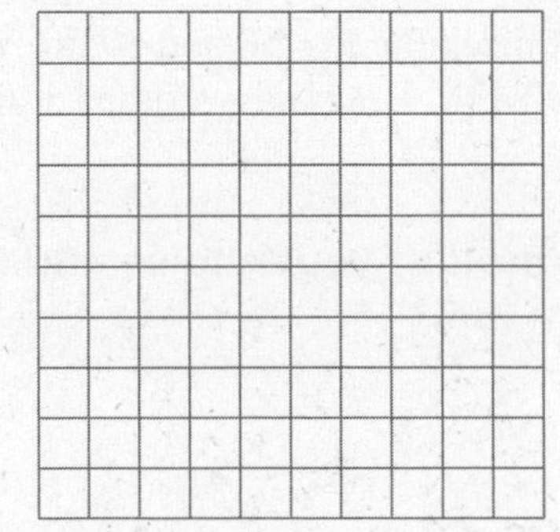

13. ____________

14. ____________

15. ____________

16. ____________

17. ____________

18. ____________

CHAPTER 7

NAME ______________________________ DATE ____________

Chapter Test B

For use after Chapter 7

1. ____________
2. ____________
3. ____________
4. ____________
5. ____________
6. ____________
7. ____________
8. ____________
9. ____________
10. ____________
11. ____________
12. ____________

Is the ordered pair a solution of the system of linear equations?

1. $-2x + 3y = 5$ $\quad (2, 3)$
   $3x + 2y = 12$

2. $2x + 5y = 23$ $\quad (-1, 5)$
   $-2x + 3y = 1$

Graph and check to solve the linear system.

3. $-2x + y = 1$
   $2x + 3y = 11$

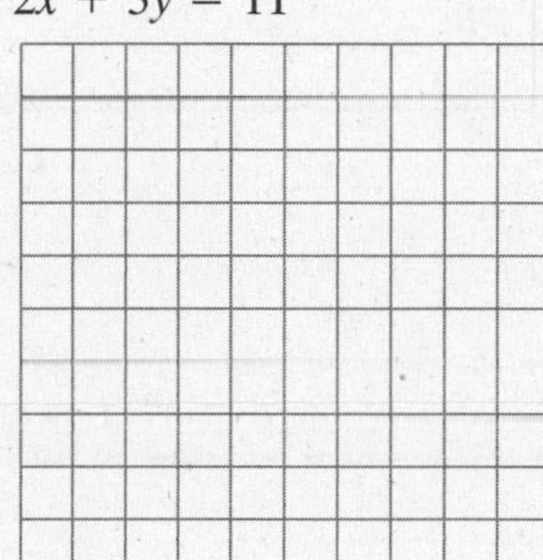

4. $-x + y = 4$
   $2x + y = 7$

Use the substitution method to solve the linear system.

5. $x + y = 4$
   $-5x + 2y = -6$

6. $3x = 9$
   $-x + 2y = 9$

7. You are selling tickets for a high school concert. Student tickets cost \$4 and general admission tickets cost \$6. You sell 450 tickets and collect \$2340. How many of each type of ticket did you sell?

Use linear combinations to solve the linear system.

8. $x + y = 3$
   $x + 2y = 6$

9. $x + y = 7$
   $y = -2x + 8$

10. A music store is selling compact discs for \$11.50 and \$7.50. You buy 12 discs and spend a total of \$106. How many compact discs that cost \$11.50 did you buy?

Solve the system using the method of your choice and tell how many solutions the system has.

11. $2x + y = 5$
    $3y = 4x - 5$

12. $4y = x + 4$
    $3x - 12y = -12$

Review and Assess

CHAPTER 7 CONTINUED

NAME ______________________________ DATE __________

Chapter Test B

For use after Chapter 7

In questions 13–16, match the system of linear inequalities with its graph.

A. $2x + 3y < 2$
$-3x + 2y \le -2$

B. $2x + 3y < 8$
$-3x + 2y \le 1$

C. $-3x + 2y > 1$
$2x + 3y \ge 2$

D. $2x + 3y \ge 8$
$-3x + 2y > -3$

13. ______________

14. ______________

15. ______________

16. ______________

17. ______________

18. ______________

13.

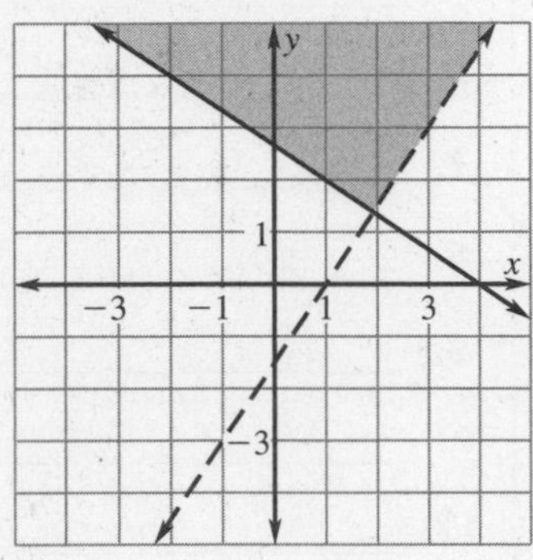

14.

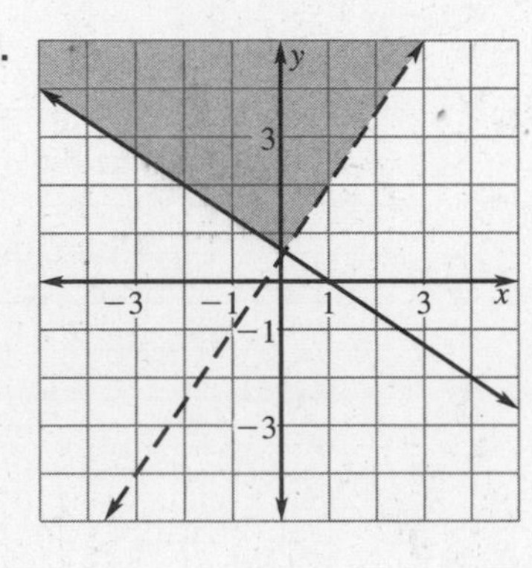

15.

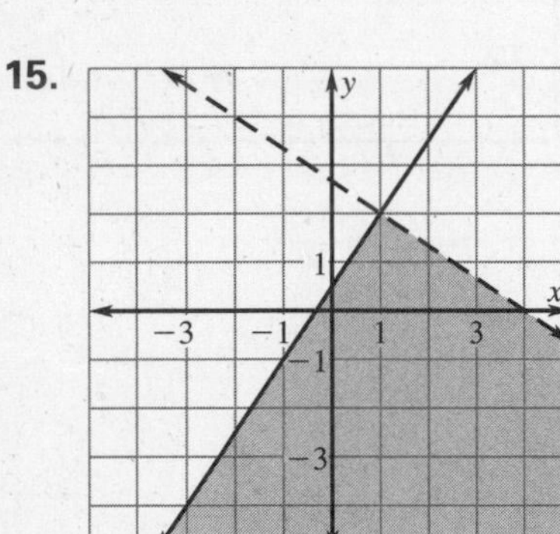

16.

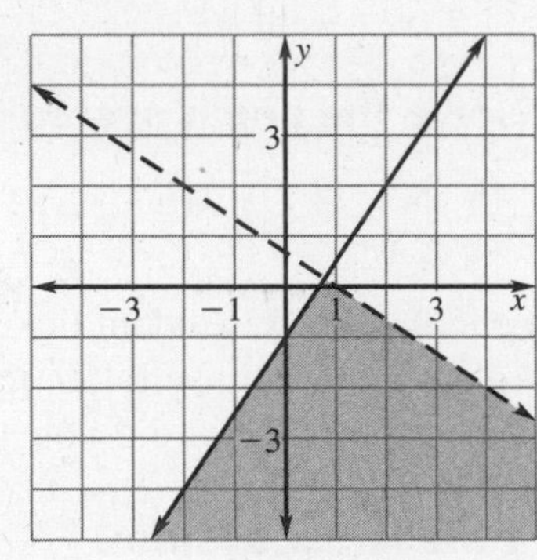

Graph the system of linear inequalities.

17. $y > x - 3$
$y \le x + 1$

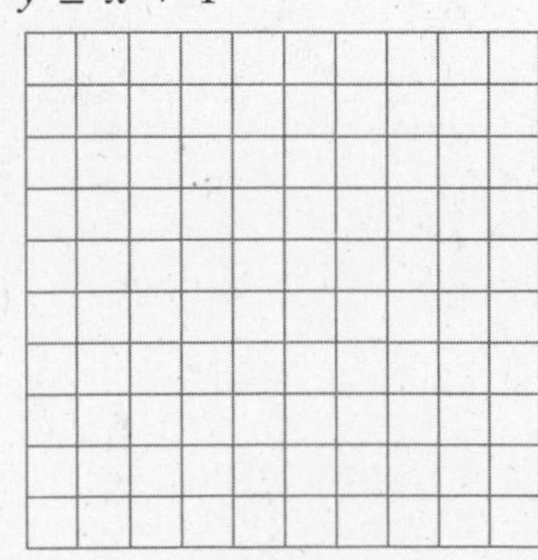

18. $y \le 2x + 3$
$y > -x + 5$

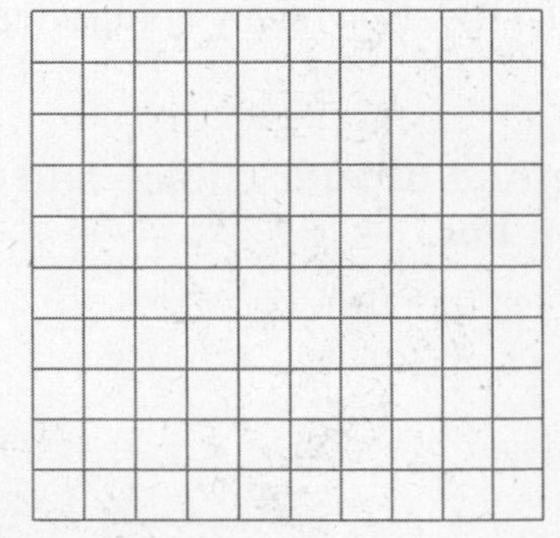

Review and Assess

CHAPTER 7

NAME ______________________ DATE __________

Chapter Test C

For use after Chapter 7

Graph and check to solve the linear system.

1. $-x + 5y = -11$
   $3x + 2y = -18$

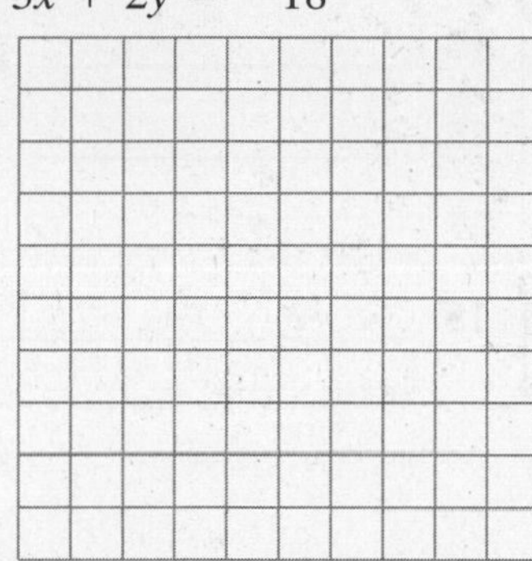

2. $2.4x + 0.8y = 0.8$
   $-0.5x + 0.25y = -1$

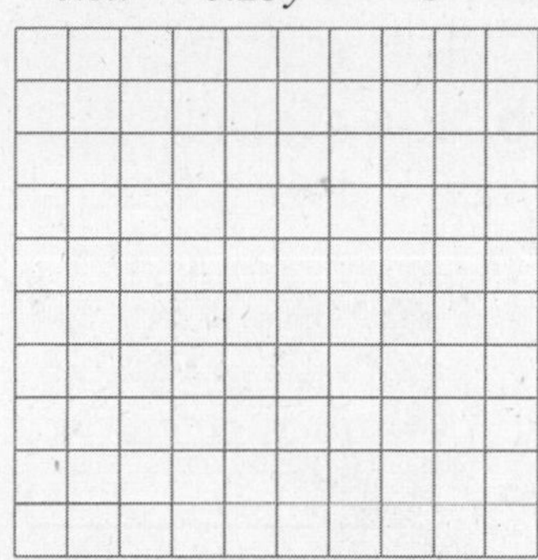

Use the substitution method to solve the linear system.

3. $-10x + y = 40$
   $-5x + 3y = -5$

4. $3x + 5y = -3$
   $-3x + y = -15$

5. One share of A stock is worth $3\frac{1}{2}$ shares of B stock. If the total value of the stock is \$9000, how much was invested in each company?

Use linear combinations to solve the linear system.

6. $4x + 5y = -2$
   $5x = 5 - 10y$

7. $3y = 16 - 2x$
   $3x + 2y = 14$

8. You call two car rental companies to find out their rental prices. Company A charges \$75 plus \$0.25 per mile and Company B charges \$80 plus \$0.30 per mile. If you are traveling 300 miles, which company gives you the better deal?

Solve the system using the method of your choice and tell how many solutions the system has.

9. $7x + 3y = -9$
   $3y = x + 15$

10. $6x - 18y = -27$
    $6y = 2x + 9$

11. $9x + y = 5$
    $-4x + 3y = -16$

12. $5y = -6x + 15$
    $12x + 10y = -5$

1. ______
2. ______
3. ______
4. ______
5. ______
6. ______
7. ______
8. ______
9. ______
10. ______
11. ______
12. ______

NAME ______________________ DATE __________

Chapter Test C

For use after Chapter 7

In Questions 13–16, match the system of linear inequalities with its graph.

A. $4x + 3y \le 13$
$-3x + 4y > 9$

B. $-3x + 4y < 5$
$4x + 3y \ge 10$

C. $-3x + 4y \le 9$
$4x + 3y > 13$

D. $4x + 3y < 10$
$-3x + 4y \ge 5$

13.

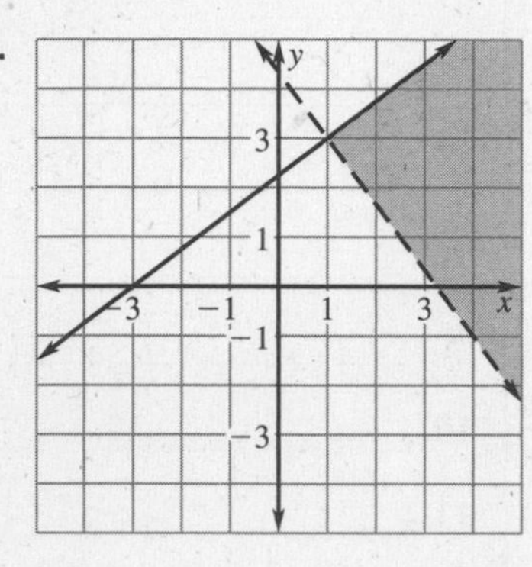

14.

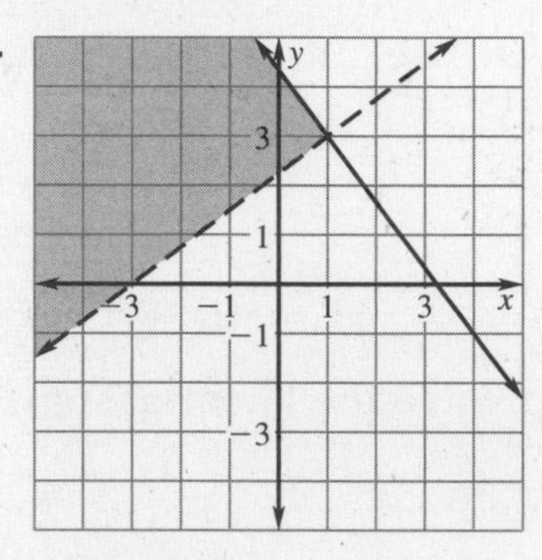

15.

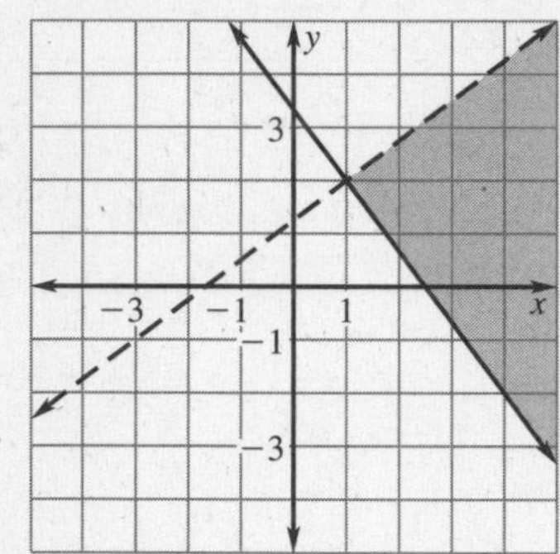

16.

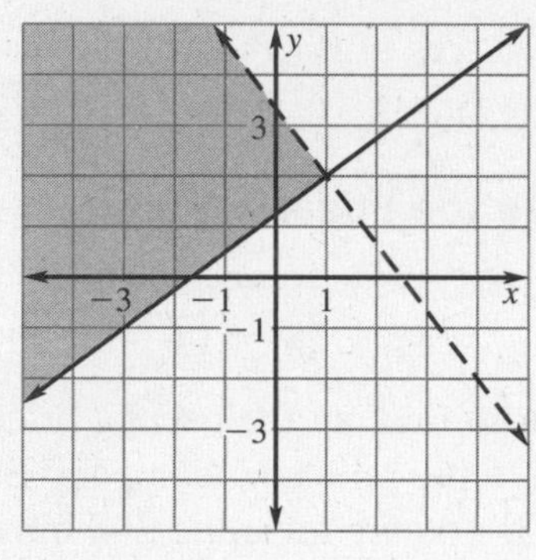

Graph the system of linear inequalities.

17. $y \le \frac{1}{2}x + \frac{5}{2}$
$y > -2x - 3$

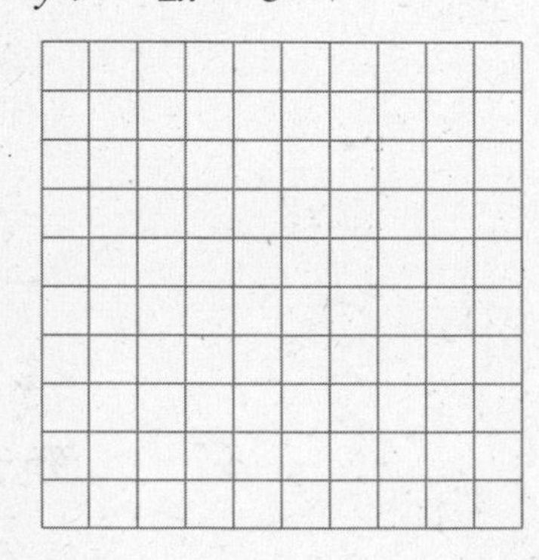

18. $y > \frac{3}{4}x + \frac{5}{4}$
$y \le -3x + \frac{5}{2}$

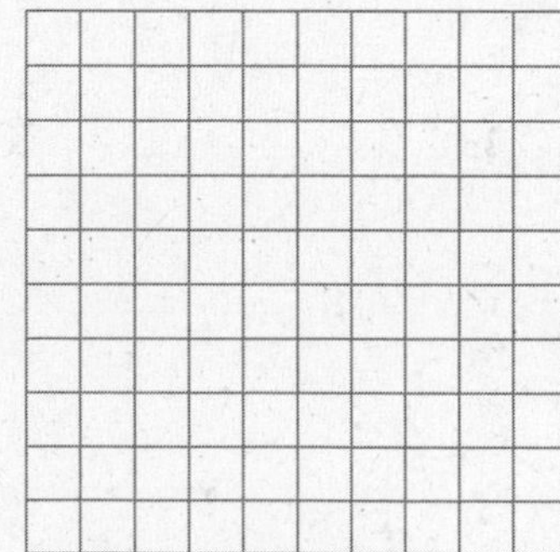

13. ____________
14. ____________
15. ____________
16. ____________
17. ____________
18. ____________

NAME ______________________ DATE __________

SAT/ACT Chapter Test

For use after Chapter 7

1. Which ordered pair is a solution of the linear system?

$4x + 3y = 5$
$-2x + 5y = 17$

Ⓐ $(2, -1)$ Ⓑ $(-1, 3)$
Ⓒ $(4, 5)$ Ⓓ $(-4, 7)$

2. If $x + 2y = 8$ and $-2x + 3y = 5$, then $x = \underline{?}$.

Ⓐ 2 Ⓑ $\frac{13}{5}$
Ⓒ 3 Ⓓ 5

3. If $4y = x + 13$ and $x + 2y = 5$, then $x + y = \underline{?}$.

Ⓐ -1 Ⓑ 2
Ⓒ 3 Ⓓ 4

4. Your teacher is giving a test worth 200 points. There is a total of 30 five-point and ten-point questions. How many five-point questions are on the test?

Ⓐ 10 Ⓑ 15
Ⓒ 20 Ⓓ 25

5. How many solutions does the linear system have? $2x - 6y = -14$; $-2x + 3y = 7$

Ⓐ none Ⓑ exactly one
Ⓒ exactly two Ⓓ infinitely many

6. At what point do the lines $5x + y = 19$ and $-x + 2y = -6$ intersect?

Ⓐ $(0, -3)$ Ⓑ $(2, 0)$
Ⓒ $(3, 4)$ Ⓓ $(4, -1)$

7. Which point is a solution of the linear system?

$2x + y = -8$
$-3x + 2y = 5$

Ⓐ $(-3, -2)$ Ⓑ $(-1, -6)$
Ⓒ $(3, 7)$ Ⓓ $(21, -50)$

8. Solve the linear system. Then choose the statement that is true about the solution of the system.

$-4x + 5y = 0$
$3x + 2y = 23$

Ⓐ The value of x is greater.
Ⓑ The value of y is greater.
Ⓒ The values of x and y are equal.
Ⓓ The relationship cannot be determined from the given information.

9. Which system of inequalities is represented by the graph?

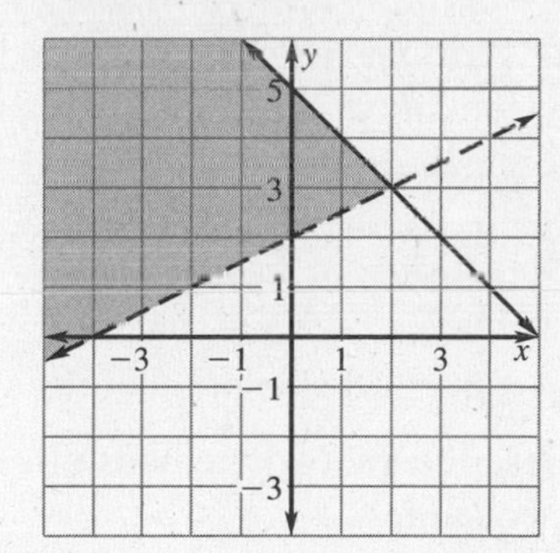

Ⓐ $y < -x + 5$, $y \geq \frac{1}{2}x + 2$ Ⓑ $y > -x + 5$, $y \leq \frac{1}{2}x + 2$
Ⓒ $y \leq -x + 5$, $y > \frac{1}{2}x + 2$ Ⓓ $y \geq -x + 5$, $y < \frac{1}{2}x + 2$

NAME ______________________ DATE ________

Alternative Assessment and Math Journal

For use after Chapter 7

JOURNAL 1. In this chapter you learned three methods to solve systems of equations. You should be able to use all three procedures and know when to choose the best method for a particular problem. In this journal you will compare the methods. (a) Write a system that is best (that is, most quickly and easily) solved by graphing. (b) Write a system that is best solved by substitution. (c) Write a system that is best solved by linear combinations. (d) Explain how to decide which method is best to use in solving a particular system. Include in your explanation a statement about why the method would be better than the others. Use the examples you wrote in parts (a)–(c) to assist you.

MULTI-STEP PROBLEM 2. Refer to the system of inequalities while doing the following.

$$\begin{cases} y \geq -3x + 5 & \text{Inequality 1} \\ 2x - 4y > 12 & \text{Inequality 2} \end{cases}$$

a. Sketch a graph of the first inequality and label it as Inequality 1.

b. On the same set of axes you used in part (a), sketch a graph of the second inequality and label it as Inequality 2.

c. In your graph, what does the overlap (or intersection) of the two half-planes represent?

d. Find the point of intersection of $y = -3x + 5$ and $2x - 4y = 12$. Is this point a solution of the system of inequalities? Explain why or why not.

3. ***Writing:*** For the questions below, find the answer to each question by looking at the graph. Explain your reasoning. Finally, check algebraically.

- Is $(6, 3)$ a solution of Inequality 1?
- Does $(-2, -4)$ satisfy Inequality 2?
- Is $(3, 0)$ a solution of the system of inequalities?

Describe the two methods of verifying a solution of a system of linear inequalities.

Alternative Assessment Rubric

For use after Chapter 7

JOURNAL SOLUTION

1. *Sample answers:*

a. $\begin{cases} y = -2x + 6 \\ y = \frac{1}{3}x - 1 \end{cases}$ **b.** $\begin{cases} x = 3y - 1 \\ 2x + 4y = 18 \end{cases}$ **c.** $\begin{cases} 3x - 8 = 28 \\ 2x + 4y = 18 \end{cases}$

d. Complete explanations should address these points:
A system is easiest to solve by graphing if both equations are in a quick-graph form such as slope-intercept. In the above example, substitution or linear combinations would be difficult since there are fractions in the problem. Substitution is best when one of the equations has an isolated variable. Linear combinations are best if both equations are in standard form.

MULTI-STEP PROBLEM SOLUTION

2. **a, b.** Check graph

c. The overlap represents the solutions of the system of inequalities.

d. $\left(\frac{16}{7}, -\frac{13}{7}\right)$, no, the point does not lie in the solution region of the system of inequalities.

3. Complete answers should addess the following points:

- (6, 3) is a solution since the point is in the shaded area for Inequality 1. Check: $3 > -13$ is true.
- $(-2, -4)$ is not a solution since the point does not lie in the solution region of the graph of Inequality 2. Check: $12 > 12$ is false.
- (3, 0) is not a solution since it is only in the shaded region for Inequality 1. Check: $0 > -4$ is true, $6 > 12$ is false.

For a graph, if a point is in the shaded area or on a solid line, the point is a solution of a system of inequalities. When checking algebraically, the point must satisfy all inequalities of the system.

MULTI-STEP PROBLEM RUBRIC

4 Students complete all parts of the questions accurately. They demonstrate an understanding that if a point is shaded or on a solid line on the graph, it will make the corresponding inequality true when substituted. Students correctly find the point of intersection.

3 Students complete the questions and explanations. Solutions may contain minor mathematical errors, such as incorrectly checking a test point or solution. Explanations may be vague.

2 Students complete questions and explanations but may make several mathematical errors. They don't find the point of intersection or may not shade the graph appropriately.

1 Student work is very incomplete. Solutions and reasoning are incorrect. Graph is missing or is completely inaccurate.

CHAPTER 7

NAME ______________________________ DATE __________

Project: Going Up!

For use with Chapter 7

OBJECTIVE **Explore how relationships among stair dimensions affect safety and comfort.**

MATERIALS inch ruler, paper, pencil, graph paper

INVESTIGATION Have you found that some stairways are more difficult to climb or easier to trip on than others? Have you thought about why this happens? Stairway safety and comfort can be influenced by the length of the riser (the vertical part between treads), the length of the tread (horizontal surface we step on), and the ratio of these lengths.

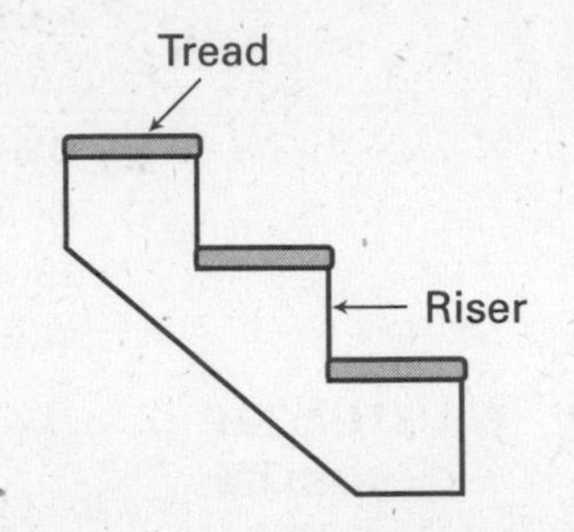

1. Make a table with the heads shown below. Collect and fill in the data for at least eight different stairways. Try for a good variety of step sizes.

Stairway location	*Tread length* (in.)	*Riser length* (in.)	*Riser/Tread ratio*	*How hard or easy to climb?*	*How safe does it feel?*

2. Based on the data you collected, what seems to be a good riser/tread ratio. Explain your decision. Did the stairways with tread lengths or riser lengths within a certain range seem safer or easier to climb? Explain.

3. Use your answers to Question 2. Design your own stairway to go up a total distance of 120 inches from one floor to the next. Determine the riser and tread lengths and the number of steps. Then make a scale drawing.

4. Below are two generally accepted rules for stairway construction. To find the riser and tread lengths that satisfy these rules, first write the relationships as four inequalities involving tread length t and riser length r. Then graph the solution of that system. Since the boundary lines are close together, use small intervals such as $\frac{1}{2}$ or $\frac{1}{4}$ along axes. (You can make the graph very large or you can use breaks in the axes and focus on the part of the graph near the solution region.)

   Rule 1: The sum of one riser length and one tread length should be from 17 inches to 18 inches.

   Rule 2: The sum of two riser lengths and one tread length should be from 24 inches to 25 inches.

5. On the same graph, plot the ordered pairs from the data you collected and from the stairway you designed. Do the stairways all satisfy the two rules?

PRESENT YOUR RESULTS Make a poster presenting your results. Include your graph and the scale drawing of the stairway you designed. Also include an analysis of the safety of the stairs for which you collected data and of the one you designed.

Review and Assess

Project: Teacher's Notes

For use with Chapter 7

GOALS

- Find and apply ratios to solve real-world problems.
- Use a system of linear inequalities to model a real-life situation.
- Solve a system of linear inequalities by graphing.

MANAGING THE PROJECT

You may wish to have students work in pairs to collect and analyze the data.

Be sure that students know to find the average width when treads are not uniform and to measure the rise to the top of the tread. For the scale model, they may find it easiest to represent 1 inch with 1 sixteenth inch or 1 millimeter.

Students may need some help in setting up their graph scales. Very small intervals are needed, but this may make it difficult to fit the y-intercepts on the graph. Several approaches are possible: If students show riser length on the y-axis and make the distance between tick marks be $\frac{1}{2}$, the y-intercepts should fit using a full sheet of quarter-inch grid paper oriented vertically. Alternatively, students can graph the boundary lines by just finding two other points on them.

Be sure that if students use a break in the x-axis, they do not graph a line by plotting the y-intercept and moving over and up to find a second point.

RUBRIC

The following rubric can be used to assess student work.

4 There is a variety of stairs chosen. Stairway design indicates the student has generalized important criteria from collected data. The student draws the graph, plots points and analyzes data correctly. The poster presents an appropriate analysis of stairway safety and makes a clear and convincing case for the safety of specific stairways. There is a discussion of whether the student's design meets the safety standards.

3 The student collects data on eight stairways, designs a stairway, draws the graph, plots points, and analyzes stairway safety. However, the student may make minor errors in measuring or in drawing the graph or may not fully address the issues when analyzing the safety of each stairway. The poster provides an analysis of stairway safety but the presentation may not be as convincing as possible.

2 The student collects data, designs a stairway, draws the graph, plots points, and analyzes stairway safety. However, work may be incomplete or show misunderstanding. For example, the student may make obvious errors in measuring, may design a stairway without attention to the riser/tread ratio, may not use the correct inequalities, or may confuse a safety issue. The poster may be unclear or may lack a significant item.

1 Data, stairway design, or graph are missing or do not show an understanding of key ideas. The poster does not give a reasonable analysis of stairway safety.

CHAPTER 7

NAME ______________________ DATE __________

Cumulative Review

For use after Chapters 1–7

Evaluate the expression. (1.3)

1. $789 - (-7^2) - 10^2$
2. $(35 \div 5) + (49 \div 7)$
3. $\dfrac{8(3 - 4) + 2}{6^2}$
4. $9 - 4^2 + (-5)^3 \div 25$

Use the distributive property to rewrite the expression without parentheses. (2.6)

5. $8(2b) - 10$
6. $-2(6 - s)$
7. $-\frac{1}{25}(30 - x)5$
8. $9.6(b - 10)$
9. $-x(4 - x) + x^2$
10. $(t + 2)t - t^2$

Solve the equation. (3.1–3.4)

11. $x + 24 = 30$
12. $9r = 89$
13. $2(q - 6) = 14$
14. $7(x - 1) = 14x - 7$
15. $\frac{3}{4}(x - 1) = 12$
16. $\frac{p}{2} - 10 = -1$

Decide whether the graphs of the two equations are parallel lines. (4.6)

17. $y = 2x - 12, 6x + 3y = 0$
18. $y = \frac{5}{6}x - 1, 5x - 6y = 36$
19. $2x + 3y = 35, y = -\frac{2}{3}x - 2$
20. $-x + 3.5y = 9, y = 0.2857x$

Find the *x*-intercept and the *y*-intercept of the graph of the equation. (4.3)

21. $6x - 2y = -7$
22. $\frac{1}{4}x = 1 - y$
23. $-4x + 5y = -8$
24. $-y = -12 + 2x$
25. $0.2x - 7 = 12y$
26. $-\frac{8}{9}x - \frac{19}{3}y = -\frac{2}{3}$

Write the equation in standard form of the line that passes through the point with the given slope. (5.6)

27. $(2, 8), m = 9$
28. $(0, 5), m = -1$
29. $(0, 0), m = \frac{3}{8}$
30. $(-6, -2), m = -\frac{9}{10}$
31. $(12, 8.56), m = 1$
32. $\left(\frac{1}{4}, \frac{3}{4}\right), m = -2$

Solve the inequality. (6.3)

33. $9 < 2x + 6 < 21$
34. $14 \geq -y + 4 \geq -30$
35. $7 + 14x < 21$ or $8x > 12$
36. $3x - 4 \geq 0$ *or* $-\frac{8}{3}x \geq 8$

Solve the equation. (6.4)

37. $|x + 6| = 10.5$
38. $|2x - 1| = 45$
39. $3|2 - x| = 9$
40. $|x - \frac{1}{8}| = 16$

NAME ______________________ DATE __________

Cumulative Review

For use after Chapters 1–7

Graph and check to solve the linear system. (7.1)

41. $y = -x + 6$
$y = x + 8$

42. $x - 4y = 15$
$x = -1$

43. $8y = -24x + 40$
$8y = 24x + 16$

44. $0.2x + y = 2$
$-x + y = 6$

Use the substitution method to solve the linear system. (7.2)

45. $y = 5x - 4y$
$x + y = 12$

46. $5g + 4h = -9$
$g - h = 0$

47. $\frac{1}{3}x + y = \frac{1}{2}$
$x - y = \frac{3}{4}$

48. $3.4x + 5.9y = 3.1$
$x + 0.1y = 1$

Use linear combinations to solve the system of linear equations. (7.3)

49. $x + y = 12$
$-3y = 4x - 10$

50. $6r + 3s = -1$
$-6r + 4s = 9$

51. $4m + 3n = 1$
$2m - 3n = 1$

52. $10x + 16y = 140$
$5x - 8y = 60$

Use the substitution method or linear combinations to solve the linear system and tell how many solutions the system has. (7.5)

53. $5x - y = 12$
$16 - 5x = -y$

54. $-4x + 6y = 18$
$9 + 2x = 3y$

55. $2x + 3y = 30$
$2x + y = 10$

56. $9.23x - 7y = 25$
$-21y = -27.69x + 75$

Graph the system of linear inequalities (7.6)

57. $x + y < 2$
$y > 5x - 1$

58. $x + y \geq 8$
$\frac{1}{5}x + 5 < 10$

59. $8x + y > 18$
$y \geq 0$
$x \geq 0$

60. $\frac{1}{2}x - 6y \geq 20$
$x - y > 5$

Review and Assess

ANSWERS

Chapter Support

Parent Guide

Chapter 7

7.1: no **7.2:** walked 12 days and jogged 7 days **7.3:** (3, −1) **7.4:** April **7.5:** infinitely many solutions **7.6:** (2, 6), (0, 2), (0, 0), (14, 0)

Prerequisite Skills Review

1. $-56r$ **2.** $\frac{1}{16}g$ **3.** $\frac{31}{10}s$ **4.** $0.08t$ **5.** 3 **6.** no solution **7.** 0.30 **8.** $\frac{10}{3}$ **9.** yes **10.** no **11.** yes **12.** no

Strategies for Reading Mathematics

1. Known: number of students in the class = 34; number of students signed up for a fall sport = 24; Unknowns: number of boys in the class (x); number of girls in the class (y); boys signed up for a fall sport = $\frac{2}{3}x$; girls signed up for sports = $\frac{3}{4}y$

2. Yes, you can write an equation to represent the total number of students in the class: $x + y = 34$. You can write an equation to represent the number of students signed up for fall sports: $\frac{2}{3}x + \frac{3}{4}y = 24$. Then you can use linear combinations or substitution to find the number of boys in the class and the number of girls in the class. **3.** Neither of the single equations you can write based on the given information has a unique solution. Neither will allow you to find either the number of boys or the number of girls in the class. Many other problems can be solved using a single equation. **4.** The answer −18 is not reasonable because there cannot be a negative number of boys in the class. The answer should be 18.

Lesson 7.1

Warm-up Exercises

1. no **2.** no **3.** yes **4.** yes

Daily Homework Quiz

1.

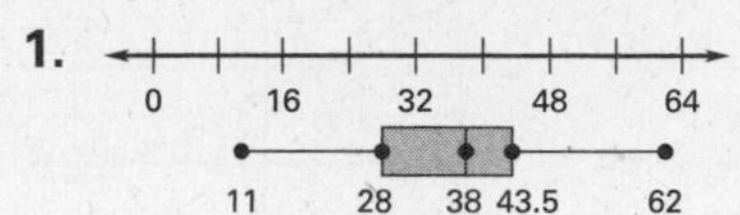

2.

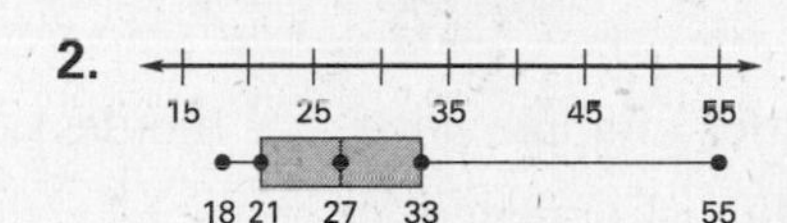

3. *Sample answer:* The lowest fourth of the values vary by only 3 mi/gal, while the highest fourth vary by 22 mi/gal.

Lesson Opener

Allow 10 minutes.

1. (3, −1) **2.** The coordinates are solutions to both equations. **3.** (2, 1) **4.** The coordinates are solutions to both equations. **5.** (−2, −4)

6. The coordintes are solutions to both equations. **7.** The coordinates of the point of intersection of two linear equations are solutions to both equations.

Practice A

1. yes; no **2.** no; no **3.** no; yes **4.** yes; no **5.** yes; no **6.** no; yes **7.** (0, 4) **8.** (1, −1) **9.** (2, 4)

10.

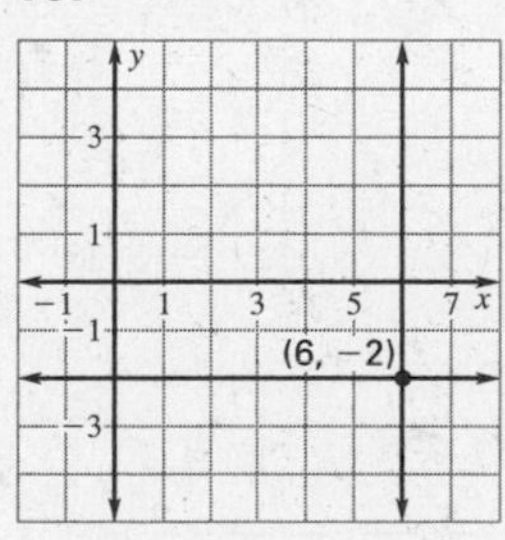

11.

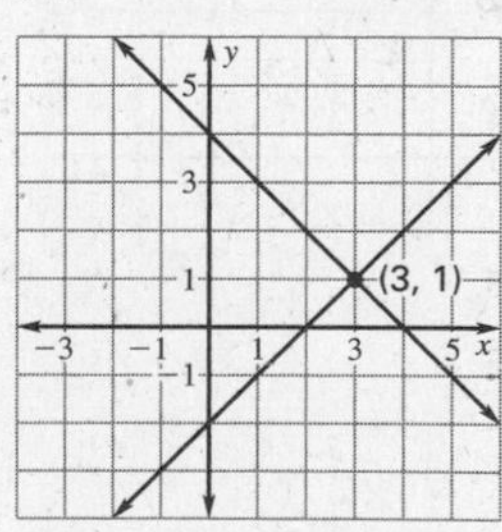

12.

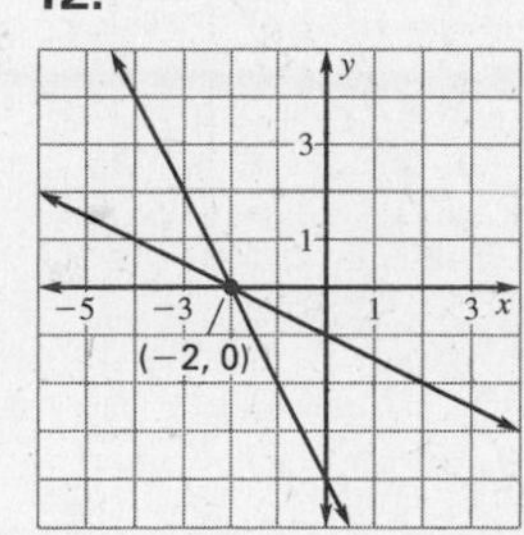

13.

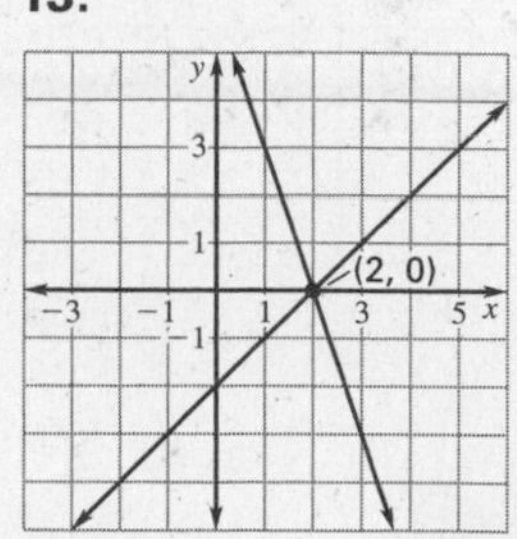

Lesson 7.1 *continued*

14.

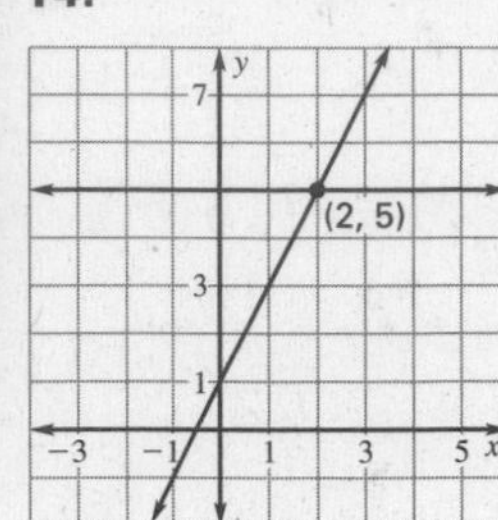

15.

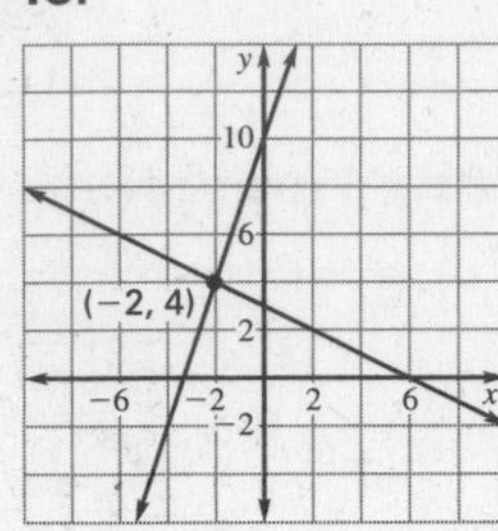

16. Number of bottles of apple juice = x (bottles)
Number of bottles of orange juice = y (bottles)
Total number of bottles = 12 (bottles)
Price per apple juice bottle = 1 (dollars)
Price per orange juice bottle = 1.5 (dollars)
Total price = 15 (dollars)
$x + y = 12$
$x + 1.5y = 15$
(6, 6); 6 1-gallon bottles of apple juice and 6 1-gallon bottles of orange juice.

Practice B

1. yes; no **2.** no; no **3.** no; yes **4.** yes; no **5.** yes; no **6.** no; no **7.** $(6, -2)$ **8.** $(-3, 3)$ **9.** $(-10, 14)$

10.

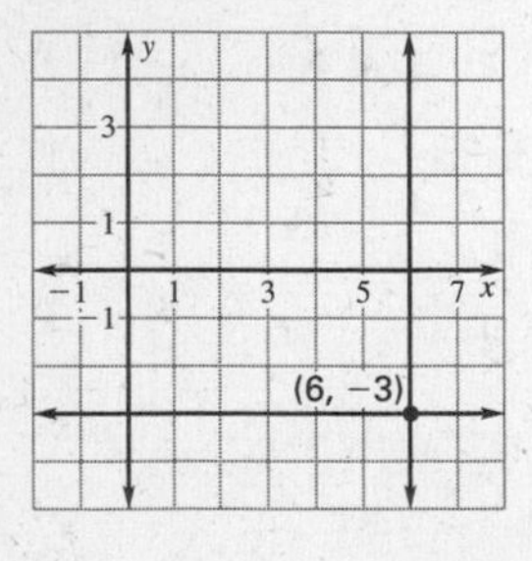

11.

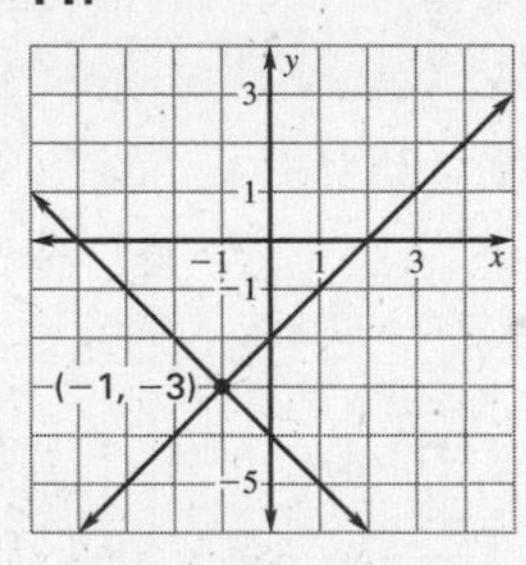

12.

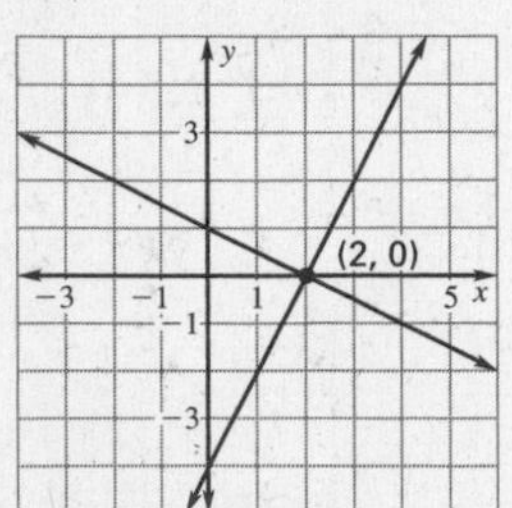

13.

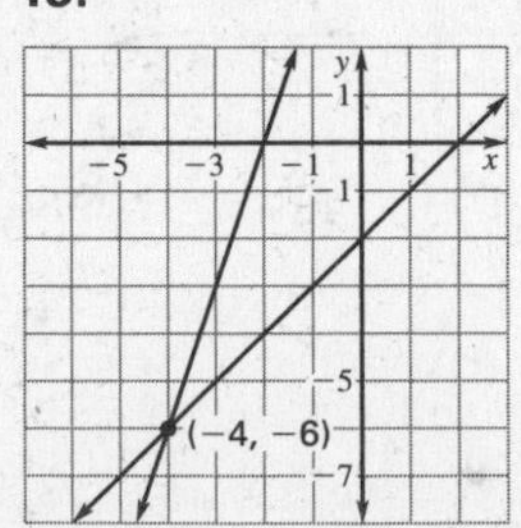

14.

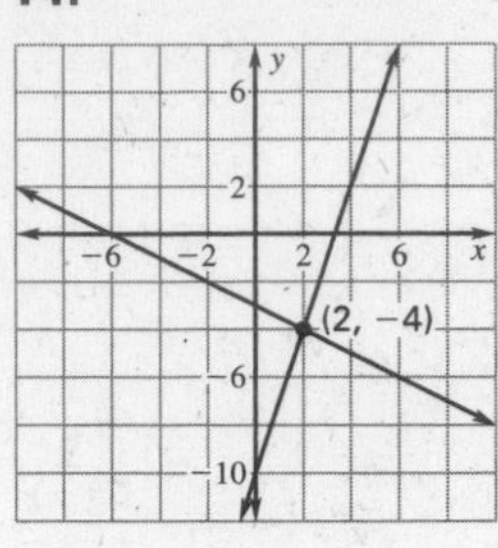

15.

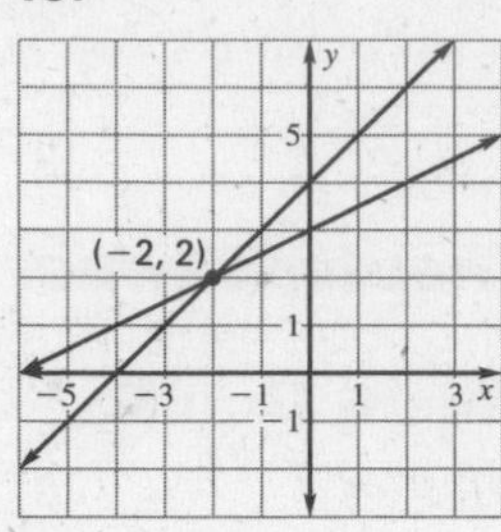

16. Number of bottles of apple juice = x (bottles)
Number of bottles of orange juice = y (bottles)
Total number of bottles = 12 (bottles)
Price per apple juice bottle = 1 (dollars)
Price per orange juice bottle = 1.75 (dollars)
Total price = 15 (dollars)
$x + y = 12$
$x + 1.75y = 15$
(8, 4); 8 1-gallon bottles of apple juice and 4 1-gallon bottles of orange juice.

17. 18; 9

Practice C

1. yes; no **2.** no; yes **3.** no; yes **4.** yes; no **5.** no; no **6.** no; yes **7.** $(4, -4)$ **8.** $\left(\frac{5}{2}, -\frac{3}{2}\right)$ **9.** $\left(\frac{7}{2}, -1\right)$

10.

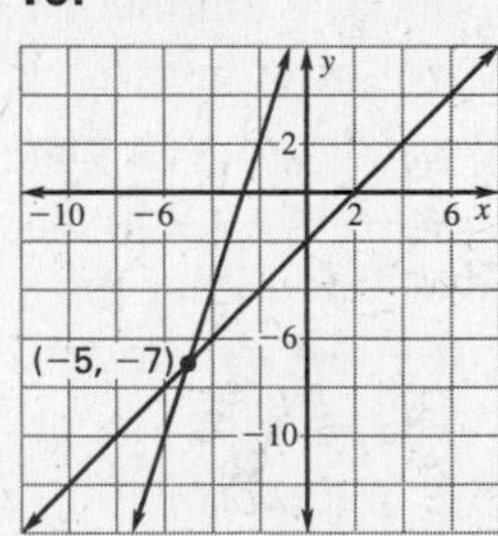

11.

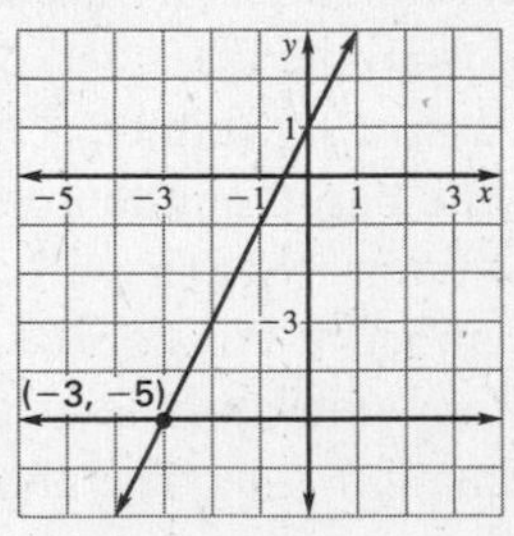

12.

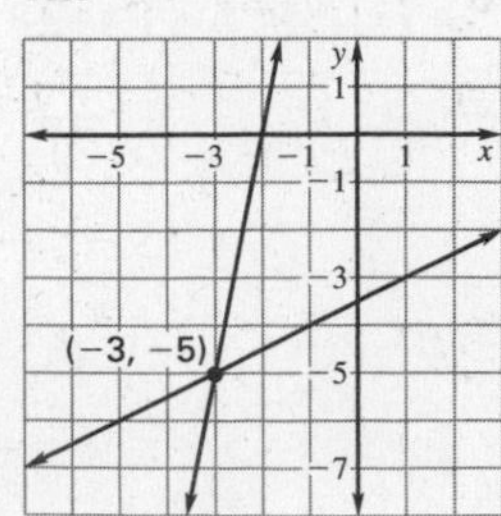

13.

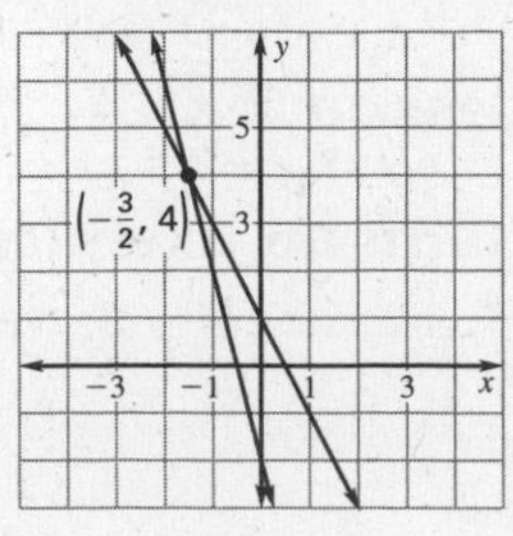

14.

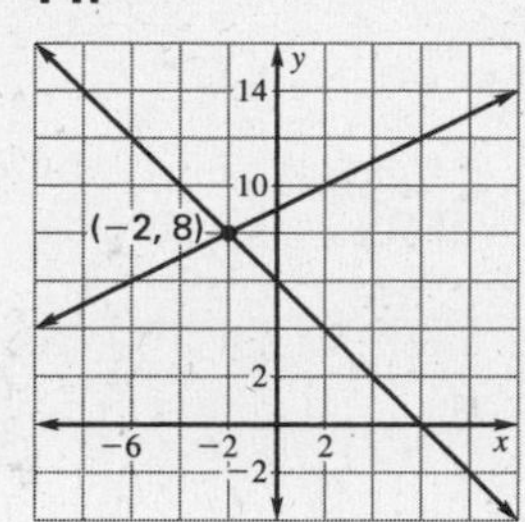

15.

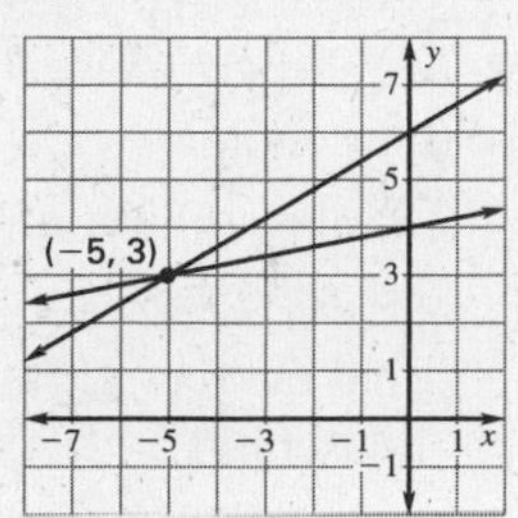

16. 8 1-gallon bottles of apple juice and 4 1-gallon bottles of orange juice.

17. \$10,000 at 5% and \$15,000 at 6%

18.

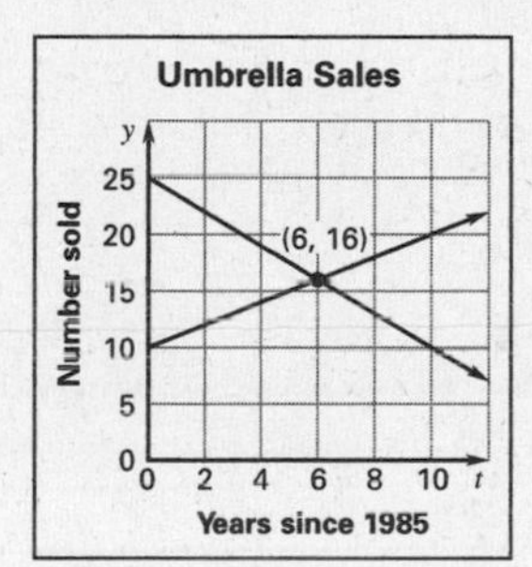

1991

Reteaching with Practice

1.

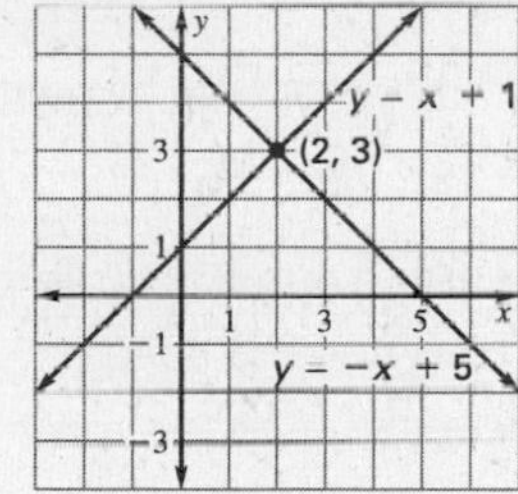

Equation 1

$$y = -x + 5$$
$$3 \stackrel{?}{=} -2 + 5$$
$$3 = 3$$

Equation 2

$$y = x + 1$$
$$3 \stackrel{?}{=} 2 + 1$$
$$3 = 3$$

2.

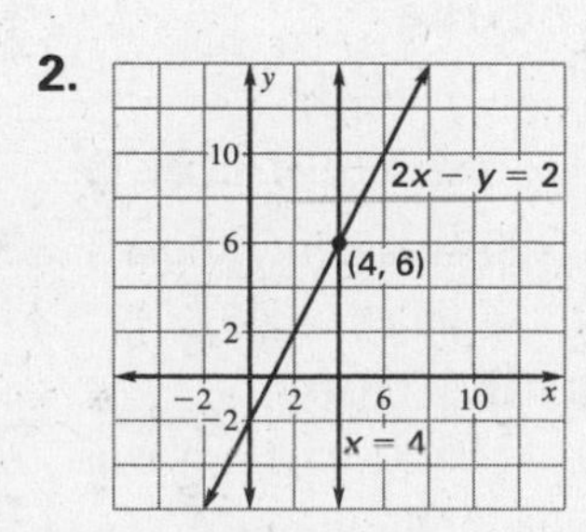

Equation 1

$$2x - y = 2$$
$$2(4) - 6 \stackrel{?}{=} 2$$
$$2 = 2$$

Equation 2

$$x = 4$$
$$4 = 4$$

3.

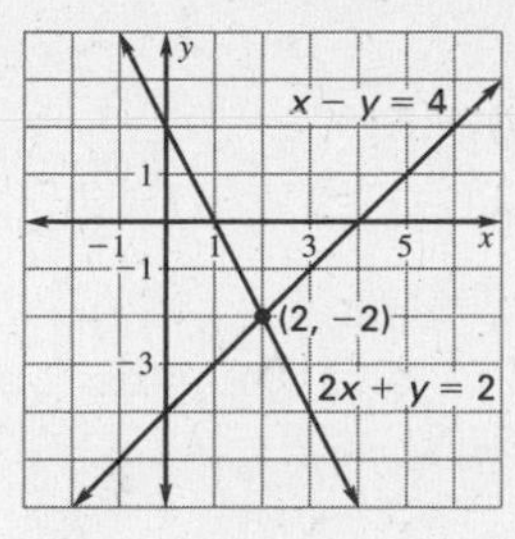

Equation 1

$$2x + y = 2$$
$$2(2) + (-2) \stackrel{?}{=} 2$$
$$2 = 2$$

Equation 2

$$x - y = 4$$
$$2 - (-2) \stackrel{?}{=} 4$$
$$4 = 4$$

4. 150 orchestra tickets **5.** 100 orchestra tickets

Real-Life Application

1. Total Papers = Current Customers + New Subscriptions · Months

Route 1 Total Papers = $36 + 3m$

Route 2 Total Papers = $20 + 5m$

2.

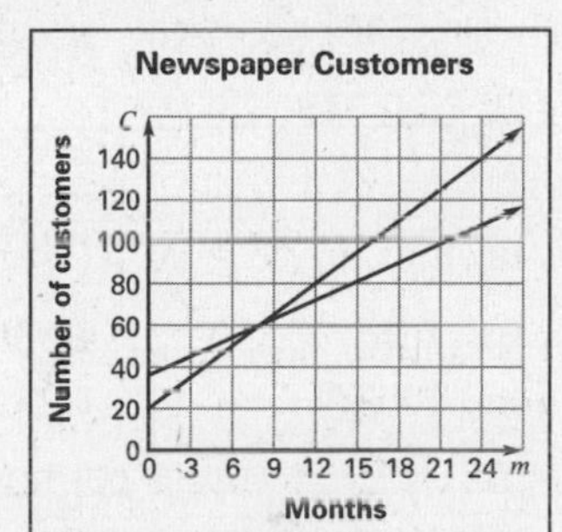

3. In 8 months **4.** Route 1 in 6 months. Route 2 in 12 months. **5.** Route 2

6. 140 Newspaper customers

7. 70 Newspaper customers

Challenge: Skills and Applications

1. no **2.** yes **3.** $y = 47t + 345$, $y = 71t + 273$ **4.** about 1998

5. $y = 547{,}725 - 25{,}195t$; $y = 493{,}559 + 27{,}146t$; $y = 237{,}612 + 12{,}831t$

6. about 1995 **7.** about 2002

Lesson 7.2

Warm-up Exercises

1. no **2.** yes **3.** no **4.** $y = -2x + 12$

5. $y = -\frac{3}{2}x + 1$

Lesson 7.2 *continued*

Daily Homework Quiz

1. yes **2.** $(-1, 1)$

3.–5. Check students' graphs.

3. $(3, -2)$ **4.** $(-1, -2)$ **5.** 15 months

Lesson Opener

Allow 10 minutes.

1–3. 1, h; 2, f; 3, e; 4, i; 5, c; 6, a; 7, d; 8, b; 9, g

4. by substitution

Practice A

1. $y = -5x + 8$ **2.** $y = 2x - 4$

3. $x = 3y + 7$ **4.** $x = -2y + 4$

5. $y = x + 3$ **6.** $x = 10y - 6$

7.–9. Answers may vary. **10.** $(2, 4)$

11. $(-1, -2)$ **12.** $(-2, 7)$ **13.** $(1, 5)$

14. $(2, -3)$ **15.** $(0, -1)$ **16.** $(3, -6)$

17. $(-2, -6)$ **18.** $(1, -3)$ **19.** $(3, 0)$

20. $(0, 0)$ **21.** $(1, -1)$

22. Number of hours brother drove $= x$ (hours)

Number of hours sister drove $= y$ (hours)

Total number of hours $= 10$ (hours)

Brother's speed $= 55$ (miles per hour)

Sister's speed $= 60$ (miles per hour)

Total number of miles $= 580$ (miles)

$x + y = 10$

$55x + 60y = 580$

(4, 6); Your brother drove 4 hours and your sister drove 6 hours.

23. 3 cm; 11 cm

Practice B

1. $y = -5x - 8$ **2.** $y = 6x - 4$

3. $x = -3y + 7$ **4.** $x = 2y - 4$

5. $y = -x - 3$ **6.** $x = 10y + 6$

7.–9. Answers may vary. **10.** $(4, 7)$

11. $(4, -7)$ **12.** $(3, -2)$ **13.** $(-5, 4)$

14. $(-4, -14)$ **15.** $(-3, 14)$ **16.** $(2, 2)$

17. $(4, -2)$ **18.** $(2, 1)$ **19.** $(-3, 2)$

20. $(6, -1)$ **21.** $(-22, 10)$

22 Number of households mow for $= x$ (households)

Number of households shovel for $= y$ (households)

Total number of households $= 10$ (households)

Earnings per household mowing $= 200$ (dollars)

Earnings per household shoveling $= 180$ (dollars)

Total earnings $= 1880$ (dollars)

$x + y = 10$

$200x + 180y = 1880$

(4, 6); You mowed for 4 households and shoveled for 6 households.

23. 5 cm; 6 cm

Practice C

1.–3. Answers may vary. **4.** $(-6, -10)$

5. $\left(-\frac{1}{2}, \frac{5}{2}\right)$ **6.** $(5, -8)$ **7.** $(-3, -7)$

8. $(-5, 1)$ **9.** $(1, 8)$ **10.** $(2, -12)$

11. $(-3, -2)$ **12.** $\left(0, \frac{1}{3}\right)$ **13.** $(2, 3)$

14. $(-1, 4)$ **15.** $(2, 2)$ **16.** $(48, -6)$

17. $\left(\frac{1}{4}, -1\right)$ **18.** $(20, 0)$ **19.** You mowed for 4 households and shoveled for 6 households.

20. $x = 12; y = 4$ **21.** 6 in., 5 in., 5 in.

22. 2 cm; 7 cm

Reteaching with Practice

1. $(-1, -2)$ **2.** $(2, 1)$ **3.** $(-1, 1)$

4. 50 shares of stock A, 150 shares of stock B

5. 25 shares of stock A, 200 shares of stock B

Interdisciplinary Application

1. $x + y = 44$ and $5x + 3y = 182$ where x is the number of frogs and y is the number of salamanders. **2.** 25 **3.** 19

4.

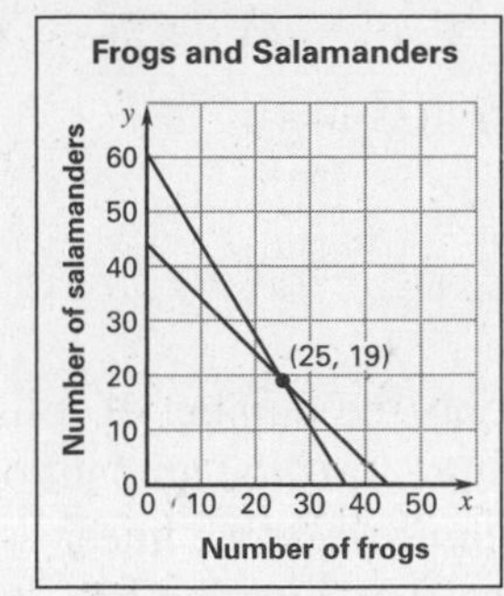

Challenge: Skills and Applications

1. $(-1, 4)$ **2.** $(3, -2)$ **3.** $(2, 9)$ and $(-2, 9)$

4. $\left(\frac{1}{3}, -1\right)$ **5.** $\frac{2}{x} + \frac{3}{y} = 8, \frac{1}{x} + \frac{6}{y} = 10$

6. $\left(2, \frac{12}{9}\right)$

7. six-penny \$0.50/lb and eight-penny \$0.75/lb

Lesson 7.3

Warm-up Exercises

1. -3 **2.** -1 **3.** $\frac{1}{2}$ **4.** $6x - 21y$

5. $-10x - 2y$

Daily Homework Quiz

1. $(5, 2)$ **2.** $(-4, 3)$ **3.** $(-3, -3)$

4. $(-7, -1)$ **5.** 43 children, 28 adults

Lesson Opener

Allow 10 minutes

1. $2x + 6y = 62$; an equation with two variables; no; there are still two different variables.

2. $-4y = -32$; an equation with one variable; yes; there is only one variable. **3.** $2x = 76$; an equation with one variable; yes; there is only one variable. **4.** $2y = 8$; an equation with one variable; yes; there is only one variable.

Practice A

1. $(2, 0)$ **2.** $(-1, 1)$ **3.** $(3, 4)$ **4.** $(6, -1)$

5. $(14, 3)$ **6.** $(2, -3)$ **7.** $(-2, -1)$ **8.** $(1, 3)$

9. $(-2, 4)$ **10.** $(3, -2)$ **11.** $(-4, 2)$

12. $(-1, 5)$ **13.** $(0, -1)$ **14.** $(-2, 2)$

15. $(-1, -4)$ **16.** $(-1, -1)$ **17.** $(5, -2)$

18. $(3, 1)$ **19.** $(-1, -2)$ **20.** $(-3, 0)$

21. $(8, 1)$

22. $y = 50 + 40x$, $y = 30 + 45x$ **23.** $(4, 210)$

24. If labor required is less than 4 hours, use Business B and if labor required is more than 4 hours, use Business A.

25. $3x + 2y = 645$, $5x + 4y = 1135$ **26.** $(155, 90)$

Practice B

1. $(9, 2)$ **2.** $(-6, -7)$ **3.** $(-2, -2)$

4. $(1, -3)$ **5.** $(-2, 2)$ **6.** $(-1, -4)$

7. $(3, -3)$ **8.** $(-5, -5)$ **9.** $(8, 3)$ **10.** $(0, 1)$

11. $(-3, 2)$ **12.** $(-2, 0)$ **13.** $(-2, 3)$

14. $(-5, -1)$ **15.** $(-3, 6)$ **16.** $(4, 6)$

17. $(-6, 0)$ **18.** $(-1, 9)$

19. $y = 50 + 36x$, $y = 35 + 39x$ **20.** $(5, 230)$

21. If labor required is less than 5 hours, use Business B and if labor required is more than 5 hours, use Business A.

22. $3x + 2y = 518$, $5x + 4y = 907$ **23.** $(129, 65.5)$

24. $x + y = 16$, $200x + 165y = 2745$ **25.** $(3, 13)$

Practice C

1. $(16, -1)$ **2.** $(10, 1)$ **3.** $(3, 7)$ **4.** $\left(\frac{1}{2}, -3\right)$

5. $(3, 6)$ **6.** $(5, 4)$ **7.** $(-1, -2)$ **8.** $(-5, 5)$

9. $(7, -2)$ **10.** $(-2, 0)$ **11.** $(-2, -3)$

12. $(0, -3)$ **13.** $(-2, 3)$ **14.** $(5, -2)$

15. $(0, 4)$ **16.** $(5, -7)$ **17.** $(6, 10)$

18. $(7, -1)$

19. $y = 50 + 39x$, $y = 32 + 45x$ **20.** $(3, 167)$

21. If labor required is less than 3 hours, use Business B and if labor required is more than 3 hours, use Business A.

22. $3x + 2y = 556.4$, $5x + 4y = 973$ **23.** $(139.8, 68.5)$

24. 3 employed at \$200 per day and 13 employed at \$165 per day

25. 240 \$28 tickets and 155 \$22 tickets

Lesson 7.3 *continued*

Reteaching with Practice

1. (3, 1) **2.** (3, −2) **3.** (−1, 3)
4. 50 smaller ads and 250 larger ads
5. 250 smaller ads and 70 larger ads

Real-Life Application

1. $-7x + 10y = 127, 1.1x + y = -7.1$
2. $x = -11, y = 5$ **3.** x and y tell you the coordinates of the island of Robinson Crusoe.

4.

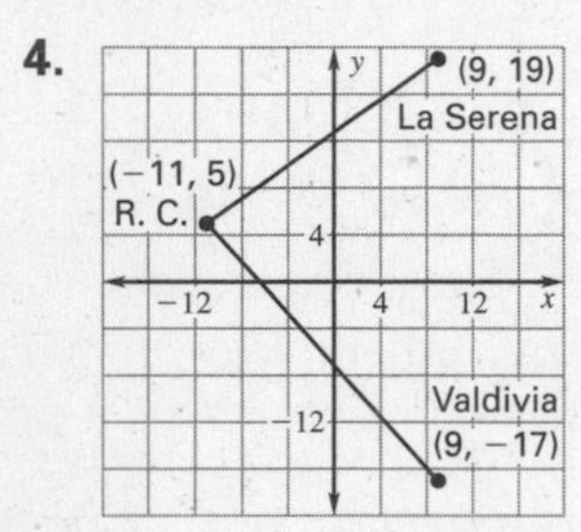

Math and History

1. (5, 7) **2.** 12 years; 10 feet
3. 22 pounds; 17 pounds

Challenge: Skills and Applications

1. (2, −3) **2.** (7, 3) **3.** $\left(-\frac{2}{a}, 5\right)$ **4.** $\left(2a, \frac{a}{2}\right)$
5. $\left(3, -\frac{5}{a}\right)$ **6.** $\left(-2, \frac{10}{b}\right)$ **7.** $\left(-\frac{13}{a}, -4\right)$
8. $\left(\frac{9}{a}, -\frac{1}{b}\right)$

Quiz 1

1. not a solution
2. (0, −2);

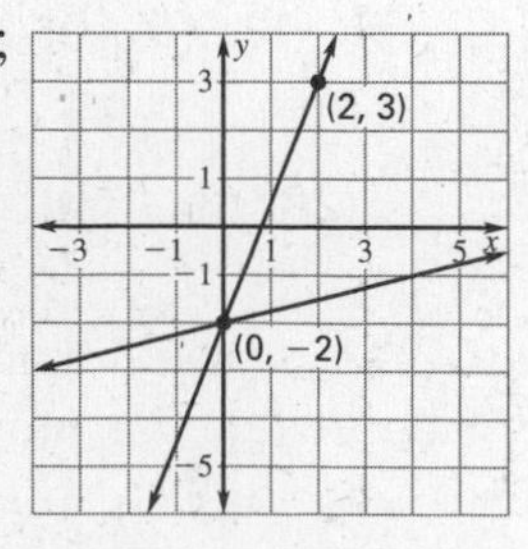

3. (7, 1)

4. (1, 4) **5.** 25 weeks

Lesson 7.4

Warm-up Exercises

1. (2, −3) **2.** (1, 1) **3.** (0, −1)

Daily Homework Quiz

1. $\left(\frac{17}{3}, 2\right)$ **2.** (−1, 5) **3.** (0, −4) **4.** $\left(-\frac{1}{2}, 5\right)$
5. (2, 3) **6.** $\left(\frac{2}{5}, -\frac{3}{5}\right)$ **7.** $\left(\frac{12}{7}, -\frac{5}{7}\right)$

Lesson Opener

Allow 10 minutes.

1. B; $x + y = 14$ represents the number of rolls sold and $6x + 8y = 92$ represents the amount of money collected. **2.** *Sample answer:* linear combinations; multiply the first equation by −6 and then add the equations. **3.** D; $x + y = 8$ represents the number of packages and $3.5x + 5y = 31$ represents the cost to ship the packages. **4.** *Sample answer:* substitution; the coefficient of x makes linear combinations messier to work with.

Practice A

1.–6. Answers may vary.

7.–12. Methods chosen and explanations may vary. **7.** (5, −4) **8.** (5, −1) **9.** (4, −2)
10. (−14, −25) **11.** (−1, −2) **12.** (−3, −2)
13. (2, 2) **14.** (1, 5) **15.** (3, 1) **16.** (3, −2)
17. (3, 1) **18.** $\left(\frac{5}{3}, \frac{2}{3}\right)$
19. $x + y = 12$
$45x + 52y = 561$
20. (9, 3)
21. $x + y = 8$
$1.6x + 5y = 23$
22. (5, 3)

23. 1 person orders the chicken dinner and 5 people order the steak dinner. **24.** The width is 4 inches and the length is 6 inches.

Practice B

1.–6. Methods chosen and explanations may vary.
1. (6, −4) **2.** (6, 2) **3.** (−1, 4) **4.** (2, 1)
5. (4, 3) **6.** (20, 30) **7.** (2, −1)
8. (−10, 3) **9.** (−1, −7) **10.** (3, 3)
11. (0, −2) **12.** (1, 2) **13.** (−4, −4)
14. (5, −2) **15.** (−2, −4) **16.** $\left(-\frac{1}{2}, 0\right)$
17. $\left(-\frac{1}{3}, 1\right)$ **18.** (−4, −6)
19. $x + y = 8$
$1.89x + 5.19y = 31.62$
20. (3, 5)

Lesson 7.4 *continued*

21. 9 right-handed gloves and 3 left-handed gloves **22.** 5 people order the chicken dinner and 1 person orders the steak dinner.

23. The width is 4 inches and the length is 6.5 inches.

Practice C

1.–6. Methods chosen and explanations may vary.
1. $(-2, 4)$ **2.** $(1, -3)$ **3.** $(-4, 6)$ **4.** $(2, 3)$
5. $(-3, 0)$ **6.** $(-5, -1)$ **7.** $(2, -1)$
8. $(-11, 3)$ **9.** $(-1, -7)$ **10.** $(3, 3)$
11. $(0, -2)$ **12.** $(2, 3)$ **13.** $(-6, -6)$
14. $(4, -1)$ **15.** $(-3, -5)$ **16.** $\left(-\frac{1}{2}, 0\right)$
17. $(-5, 2)$ **18.** $(-6, -9)$ **19.** $\left(-\frac{10}{7}, -\frac{6}{7}\right)$
20. $(-4, -19)$ **21.** $(6, 8)$
22. $x + y = 8$
$2.19x + 5.89y = 28.62$ **23.** $(5, 3)$

24. 2 people order the chicken dinner and 4 people order the steak dinner.

25. The width is 4.5 inches and the length is 6.5 inches. **26.** 3 hours; 2 hours

Reteaching with Practice

1. Sample answers are given. **a.** Substitution method because the coefficient of x is 1 in Equation 2. **b.** Linear combination method because none of the variables has a coefficient of 1 or -1. **c.** Substitution method because the coefficient of x is 1 in Equation 1.

2. Sugar: \$.35 per pound, flour: \$.25 per pound

Cooperative Learning Activity

1. Todd and Shelli will have spent the same amount for CDs when they have purchased 57 CDs at the cost of about \$797.43.

2. It would be less expensive to buy CDs from the music store for purchases above the initial 57 CDs. Otherwise, it would be a better deal to buy CDs through the mail.

Interdisciplinary Application

1. $x =$ number of trumpets; $y =$ number of trombones; $x + y = 27$

2. $350x + 475y = 10{,}950$

3. Answers may vary. *Sample answer:* Use substitution to find $x = 15$ and $y = 12$.

4. Answers may vary. *Sample answer:* Use linear combinations to find $x = 15$ and $y = 12$.

5. 15 trumpets, 12 trombones

Challenge: Skills and Applications

1. *Sample answer:* substitution; it is easy to solve the first equation for y or the second equation for x; $(a, -2a)$ **2.** *Sample answer:* linear combinations; it is easier to multiply one equation by 2 than it is to solve for either x or y in either equation, $\left(\frac{8}{a}, -\frac{6}{b}\right)$ **3.** $p = u + 10d$, $q = d + 10u$; $p - q = -9u + 9d$, which is divisible by 9.

4. $\frac{96}{x} + \frac{160}{y} = 9$, $\frac{144}{x} + \frac{160}{y} = 11$

5. $u = \frac{1}{24}$, $v = \frac{1}{32}$ **6.** 24 mi/gal, 32 mi/gal

Lesson 7.5

Warm-up Exercises

1. $y = -x + 12$; $-1, 12$ **2.** $y = \frac{1}{3}x + 1$; $\frac{1}{3}, 1$
3. $y = \frac{2}{3}x - \frac{5}{2}$; $\frac{2}{3}, -\frac{5}{2}$ **4.** $y = \frac{1}{5}x - 2$; $\frac{1}{5}, -2$

Daily Homework Quiz

1. $(-1, -3)$ **2.** $(2, -5)$ **3.** $\left(\frac{5}{3}, 2\right)$
4. $(2, 1)$ **5.** $\left(\frac{1}{6}, \frac{5}{6}\right)$ **6.** $(6, -7)$
7. 16 or more CDs

Lesson Opener

Allow 15 minutes.

1. The lines intersect and there is one solution. **2.** The lines are the same and there are infinitely many solutions. **3.** The lines are parallel and there are no solutions. **4.** The lines are parallel and there are no solutions. **5.** The lines intersect and there is one solution. **6.** The lines are the same and there are infinitely many solutions.

7. The lines are the same and there are infinitely many solutions. **8.** The lines intersect and there is one solution. **9.** 1, 5, and 8; The lines intersect in one point. **10.** 3 and 4; The lines are parallel. **11.** 2, 6, and 7; The lines are the same.

Lesson 7.5 *continued*

12. If the lines intersect at one point, the system has one solution. If the lines are parallel (but not the same), the system has no solutions. If the lines are the same, the system has infinitely many solutions.

Graphing Calculator Activity

1. a.

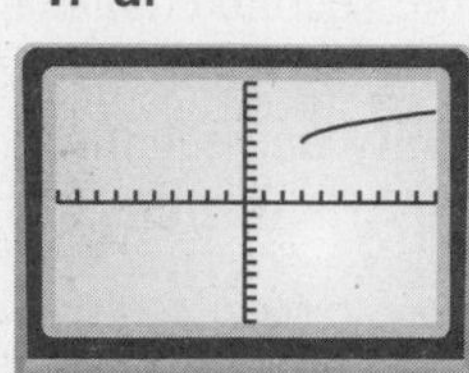

b.

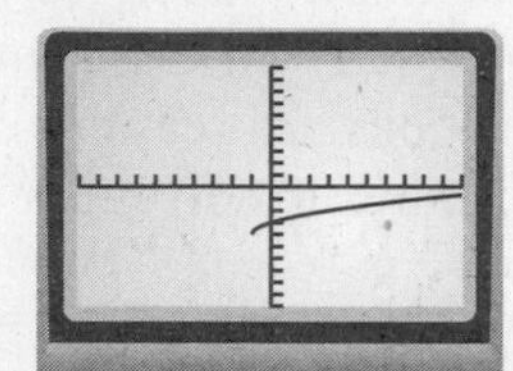

c.

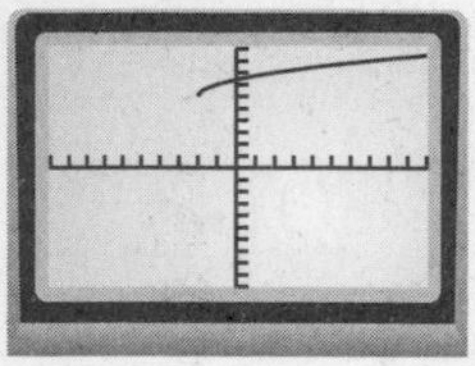

2. right **3.** left **4.** up **5.** down

Practice A

1. E; exactly one solution **2.** A; no solution
3. C; infinitely many solutions
4. B; exactly one solution
5. F; infinitely many solutions
6. D; no solution **7.** no solution
8. no solution **9.** no solution
10. no solution **11.** infinitely many solutions
12. exactly one solution $(-1, -1)$
13. no solution **14.** infinitely many solutions
15. no solution **16.** no solution
17. exactly one solution $(0, -3)$
18. infinitely many solutions
19.

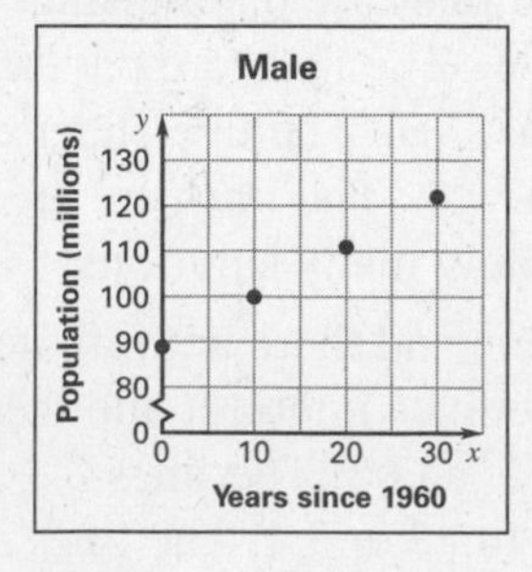

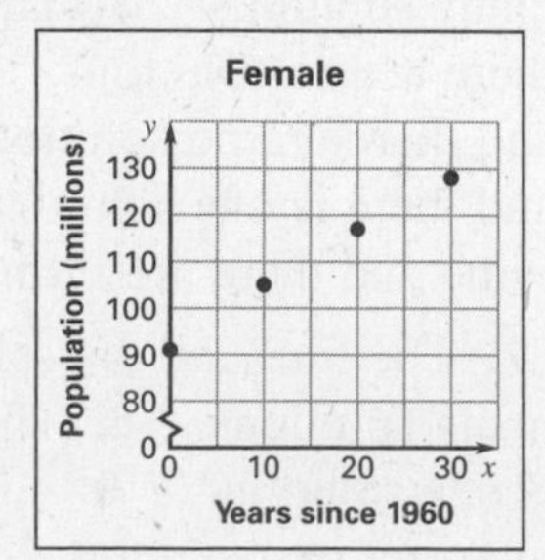

Male: $y = 1.1x + 89$
Female: $y = 1.2x + 92$

20. No; No, because the female population is growing at a faster rate.

Practice B

1. E; exactly one solution **2.** A; no solution
3. C; infinitely many solutions
4. B; exactly one solution
5. F; infinitely many solutions. **6.** D; no solution **7.** no solution **8.** infinitely many solutions **9.** exactly one solution $(2, -5)$
10. exactly one solution $\left(-\frac{1}{3}, 1\right)$
11. infinitely many solutions **12.** no solution
13. no solution **14.** no solution
15. no solution **16.** no solution
17. exactly one solution $(3, -1)$
18. no solution
19.

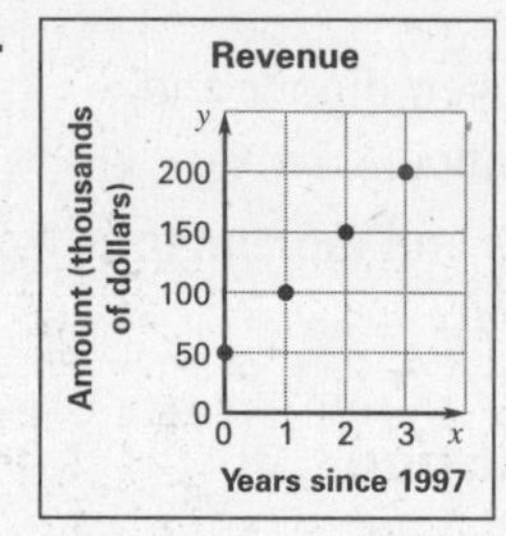

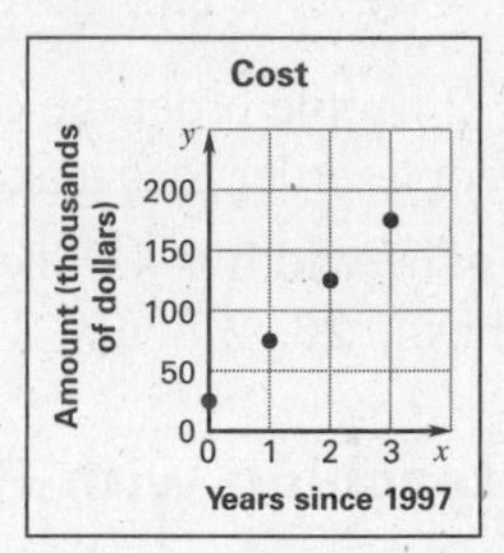

Revenue: $y = 50x + 50$
Cost: $y = 50x + 25$

20. The profit was constant.

Practice C

1. B; exactly one solution **2.** A; no solution
3. C; infinitely many solutions **4.** no solution
5. infinitely many solutions
6. infinitely many solutions
7. exactly one solution $(-1, -1)$
8. no solution **9.** no solution
10. no solution **11.** no solution
12. infinitely many solutions **13.** no solution
14. exactly one solution $(0, -2)$
15. infinitely many solutions

Lesson 7.5 *continued*

16.

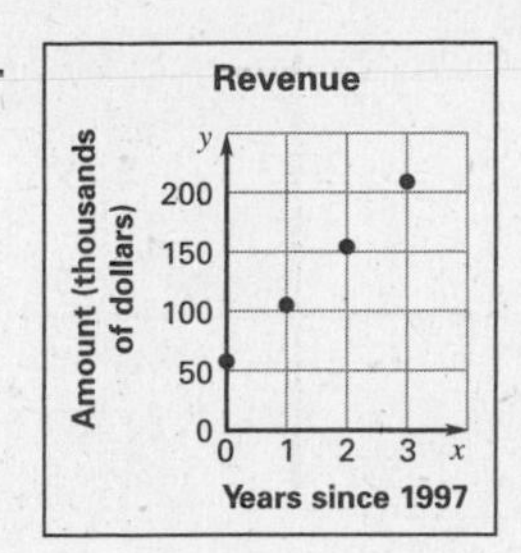

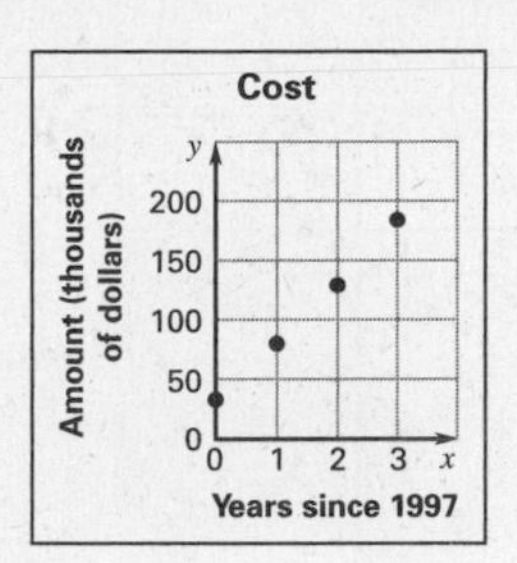

Revenue: $y = 50.2x + 56.2$

Cost: $y = 50.2x + 31.2$

17. The profit was constant, $25,000 per year.

18. No, they are equivalent equations.

Reteaching with Practice

1.–6. Methods may vary. Method (a) is substitution or linear combination. Method (b) is graphing.

1. a. None; because $-4 = -1$ is a false statement, the system has no solution. **b.** None; because the lines are parallel, the system has no solution.

2. a. None; because $5 = \frac{7}{3}$ is a false statement, the system has no solution. **b.** None; because the lines are parallel, the system has no solution.

3. a. None; because $1 = -1$ is a false statement, the system has no solution. **b.** None; because the lines are parallel, the system has no solution.

4. a. Infinitely many; because $0 = 0$ is a true statement, the system has infinitely many solutions.

b. Infinitely many; because the lines coincide, the system has infinitely many solutions.

5. a. Infinitely many; because $0 = 0$ is a true statement, the system has infinitely many solutions.

b. Infinitely many; because the lines coincide, the system has infinitely many solutions.

6. a. Infinitely many; because $0 = 0$ is a true statement, the system has infinitely many solutions.

b. Infinitely many; because the lines coincide, the system has infinitely many solutions.

7. The price of one sketchpad is $3.

Real-Life Application

1.

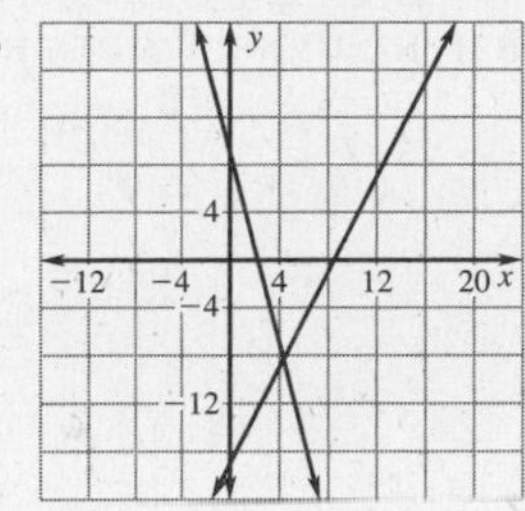

Yes, there is a possibility because the two lines intersect.

2. The two equations represent the same line. If the cubs are walking toward you, then yes, you will see them. If the cubs are walking away from you, then no, you will not see them.

3. No, you will not cross the stream. The two lines are parallel.

Challenge: Skills and Applications

1. $\frac{2}{5}$ **2.** none **3.** all k not equal to $\frac{2}{5}$

4. no; yes **5.** When $k = 2$, the system has infinitely many solutions; when k does not equal 2, it has no solution. **6.** $ad - bc$

7. *Sample answer:* $ad = bc$

Lesson 7.6

Warm-up Exercises

1. solid **2.** dashed **3.** $4x + 6y \le 9$

Daily Homework Quiz

1. no solution **2.** 1 solution

3. 1 solution, $(-1, 2)$

4. infinitely many solutions

5. no; the system $4h + 4s = 8$, $6h + 6s = 12$ has infinitely many solutions.

Lesson Opener

Allow 10 minutes.

1–2.

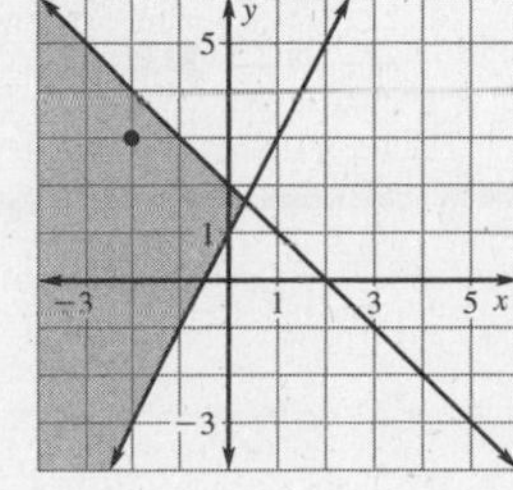

3. $y \ge 2x + 1$

Lesson 7.6 *continued*

4. $y \le -x + 2$ **5.** The shaded region is the intersection of the regions that are the solutions of the two inequalities.

6–7.

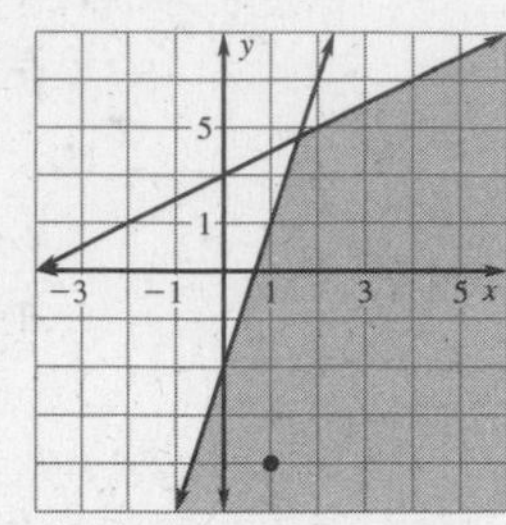

8. $y \le 3x - 2$

9. $y \le \frac{1}{2}x + 2$ **10.** The shaded region is the intersection of the regions that are the solutions of the two inequalities.

Graphing Calculator Activity

1. a.– c. Check graphs.

2. a. $\ge, \le$ **b.** $\ge, \le$ **c.** $\le, \le$

Practice A

1. C **2.** A **3.** B **4.** $y \ge x - 3$, $y \ge -2x + 6$

5. $y < \frac{1}{2}x + 3$, $y \ge \frac{1}{2}x - 2$ **6.** $y < -\frac{1}{2}x + 1$, $y < 2x + 1$

7.

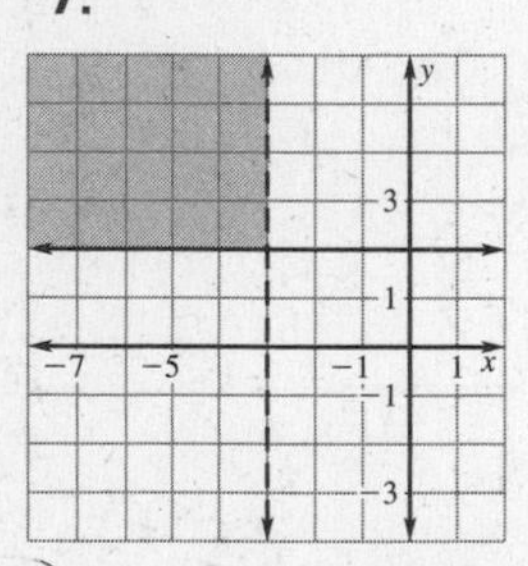

8.

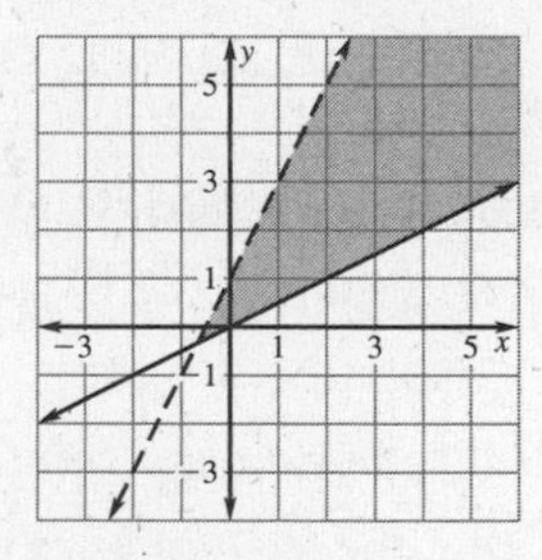

9.

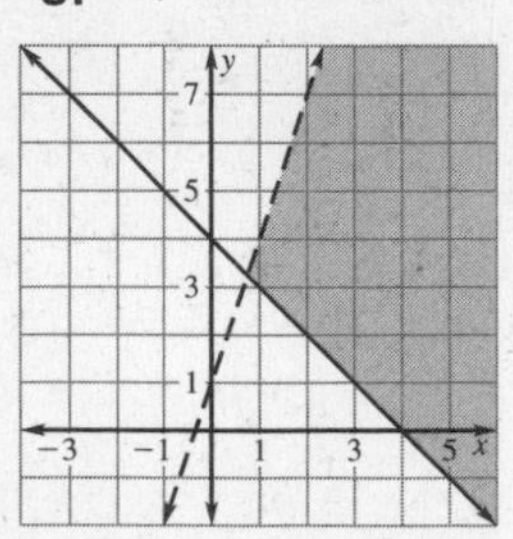

10.

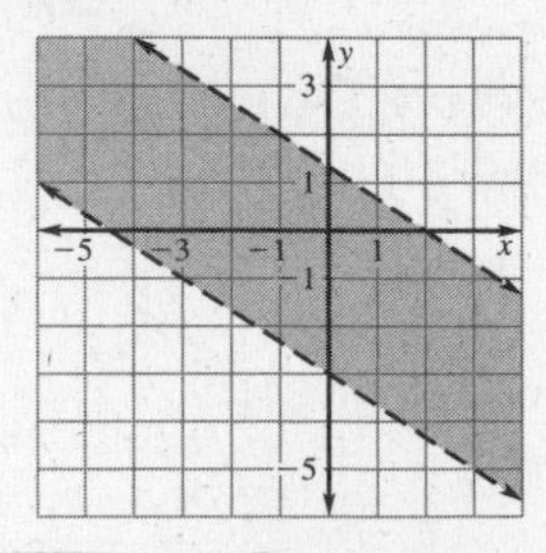

11.

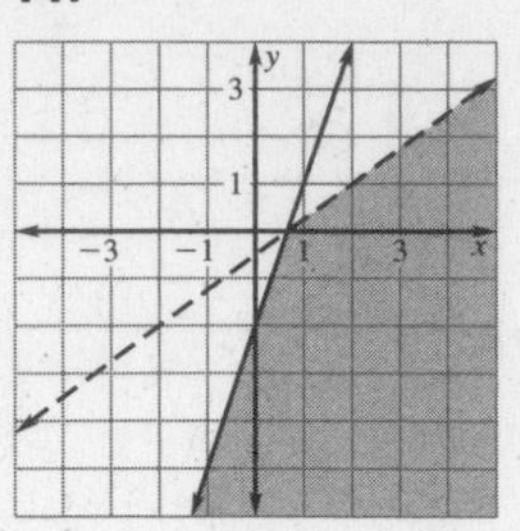

12.

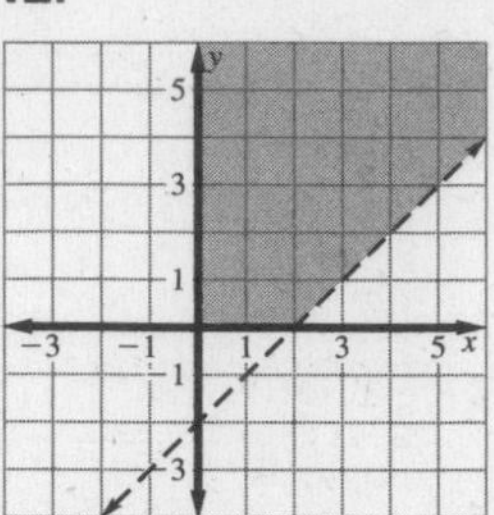

13. $x + y \ge 3$
$x + y \le 7$
$x \ge 0$
$y \ge 0$

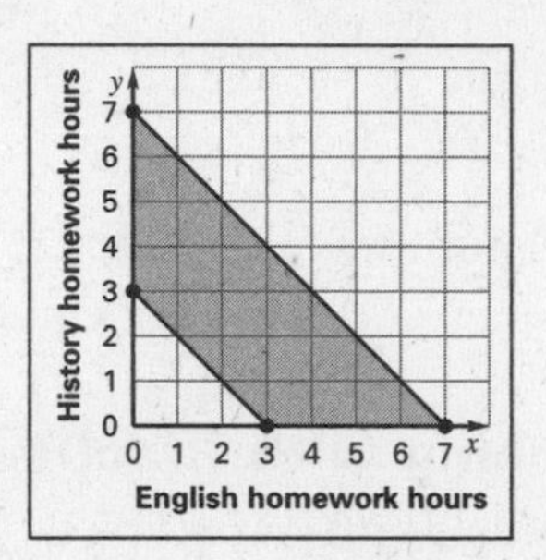

14. $100x + 150y \le 1200$
$x + y \le 10$
$x \ge 0$
$y \ge 0$

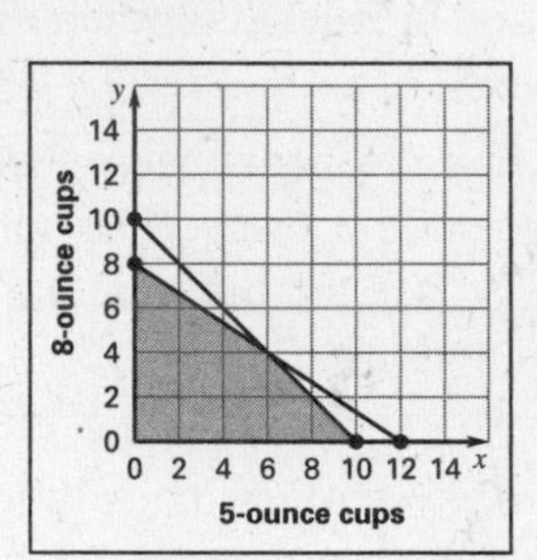

15. $y \le \frac{1}{2}x + 3$
$y \ge \frac{1}{2}x - 1$
$x \le 2$
$x \ge -2$

Practice B

1. C **2.** A **3.** B **4.** $y > 2x + 7$, $y \le -2x - 1$

5. $y \le -\frac{1}{2}x + 3$, $y \ge -\frac{1}{2}x - 2$ **6.** $y < -\frac{1}{2}x + 4$, $y < 2x + 1$

7.

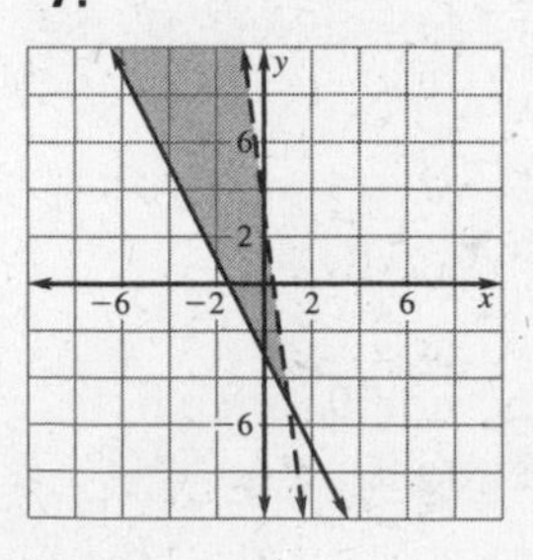

8.

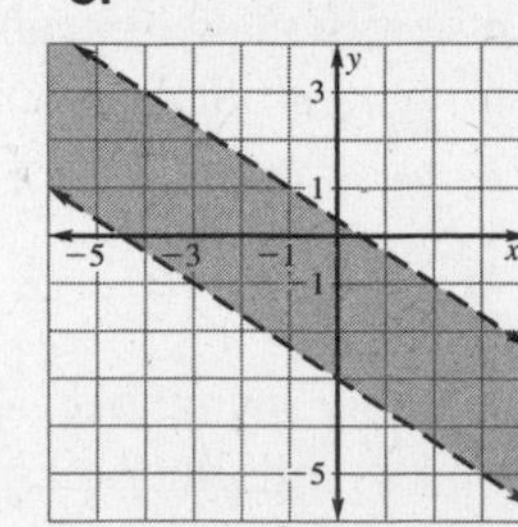

9.

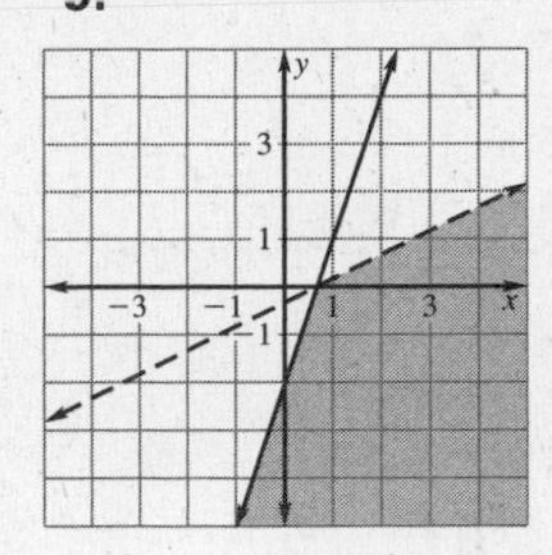

10.

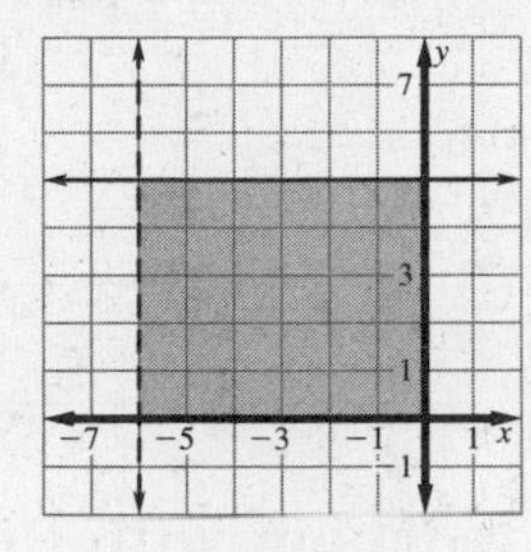

11.

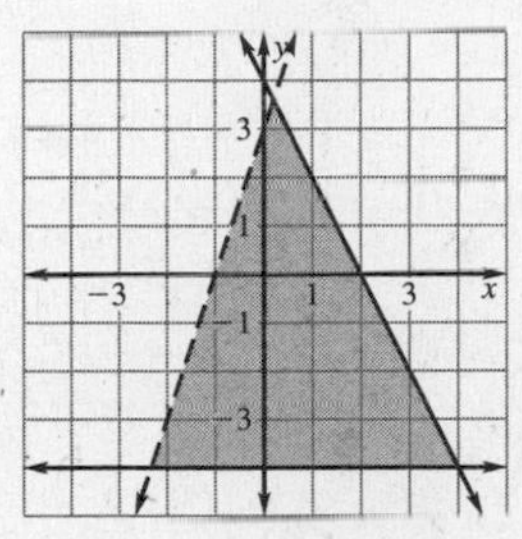

12.

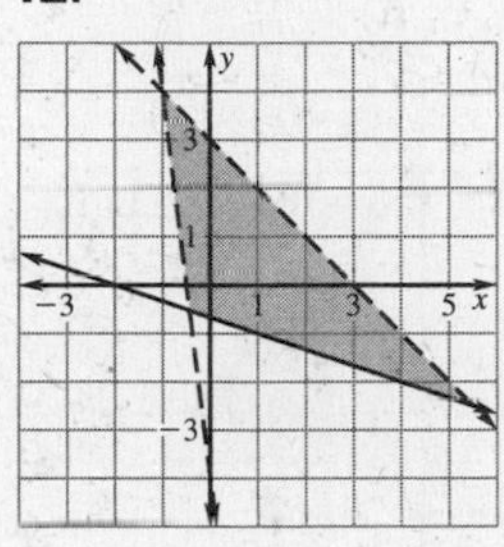

13.

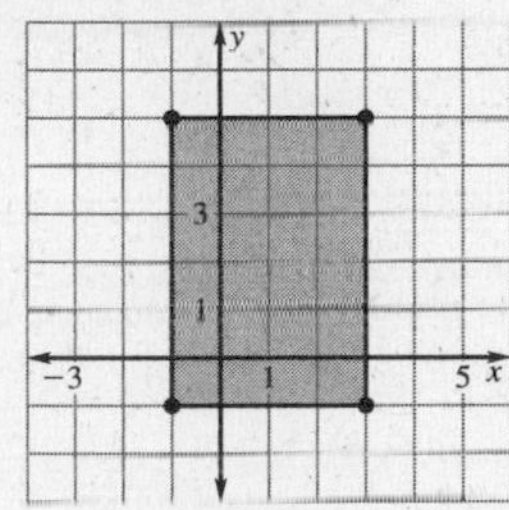

$x \geq -1$
$x \leq 3$
$y \leq 5$
$y \geq -1$

14.

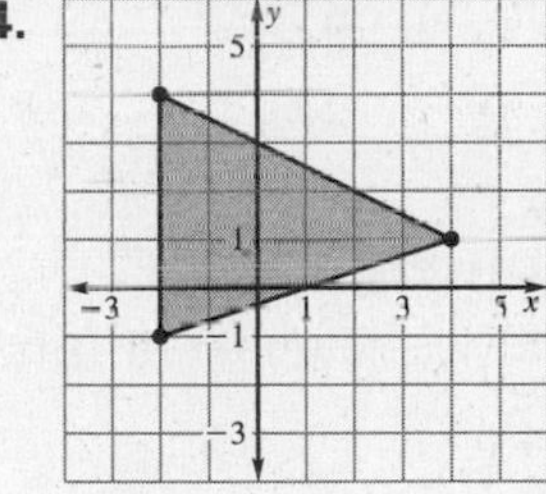

$x \geq -2$
$y \leq -\frac{1}{2}x + 3$
$y \geq \frac{1}{3}x - \frac{1}{3}$

15.

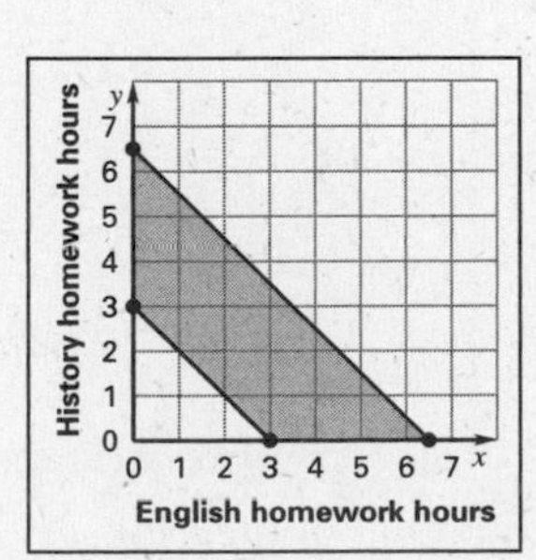

$x + y \geq 3$
$x + y \leq \frac{13}{2}$
$x \geq 0$
$y \geq 0$

16.

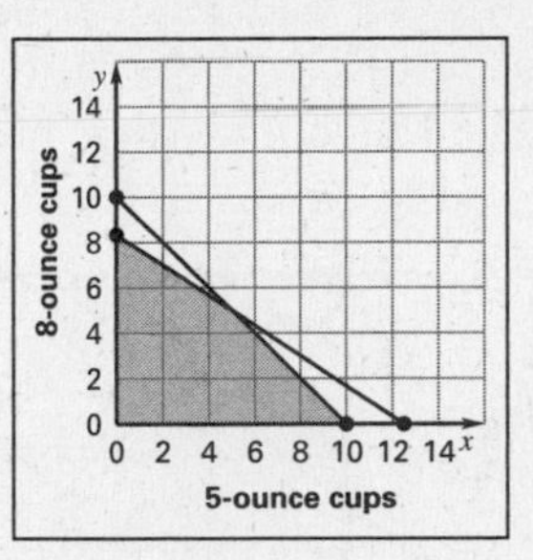

$100x + 150y \leq 1250$
$x + y \leq 10$
$x \geq 0$
$y \geq 0$

Practice C

1. $y \geq -\frac{1}{2}x - 5$
$y > 3x + 2$

2. $y < \frac{2}{3}x - 1$
$y > \frac{2}{3}x - 4$

3. $y \geq 2x + 1$
$y \leq -\frac{1}{2}x + 4$

4.

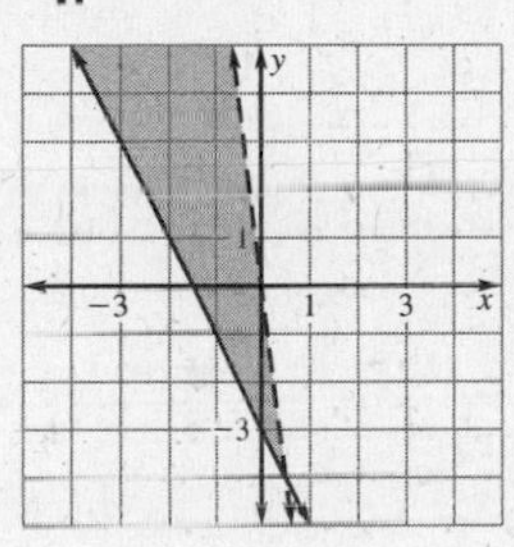

5.

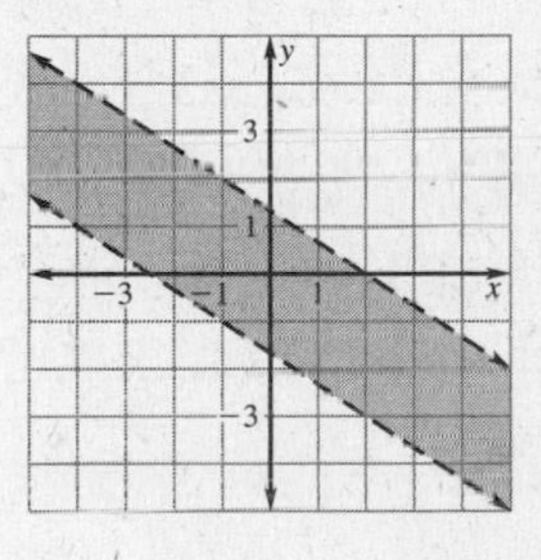

6.

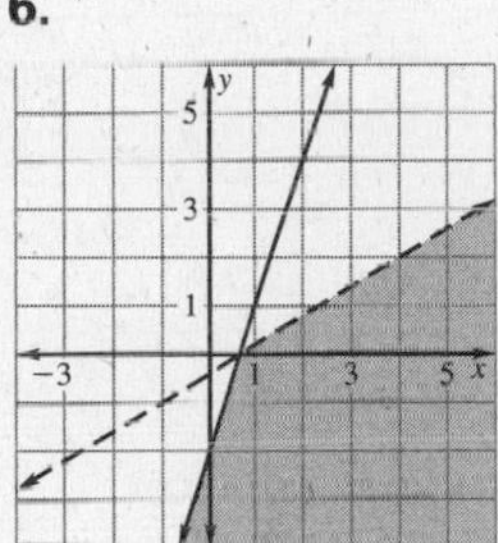

7.

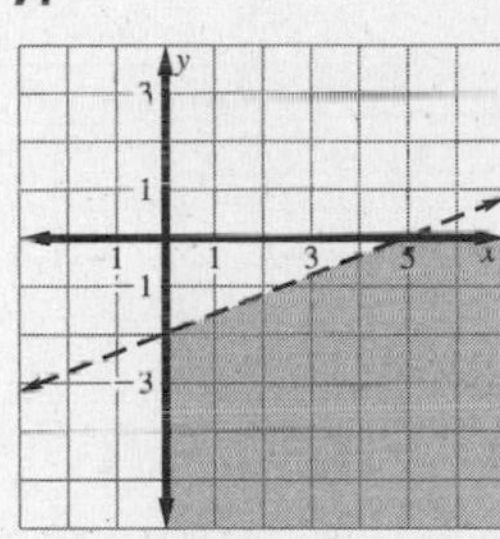

8.

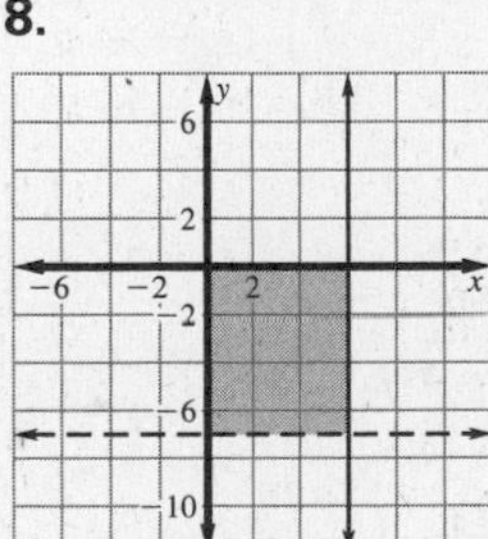

9.

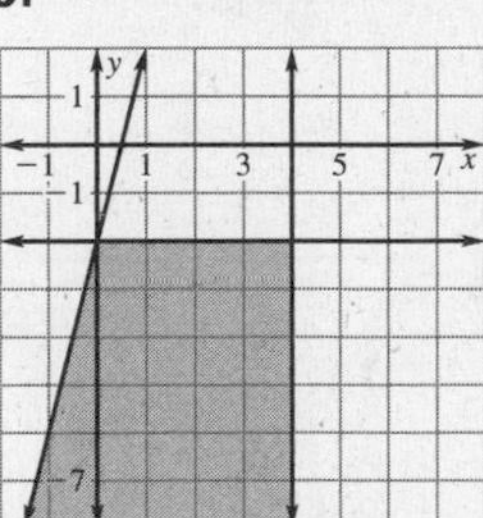

Lesson 7.6 *continued*

10.

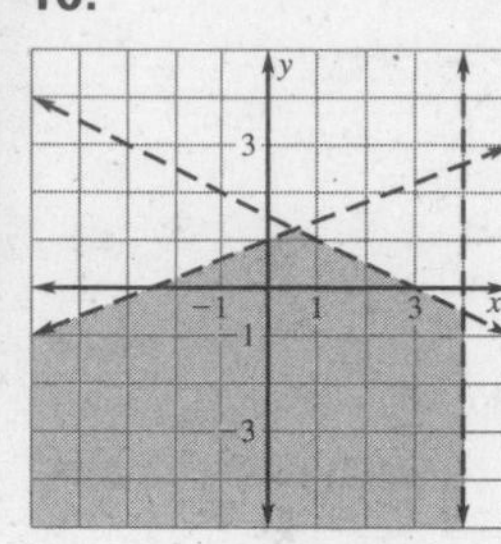

11.

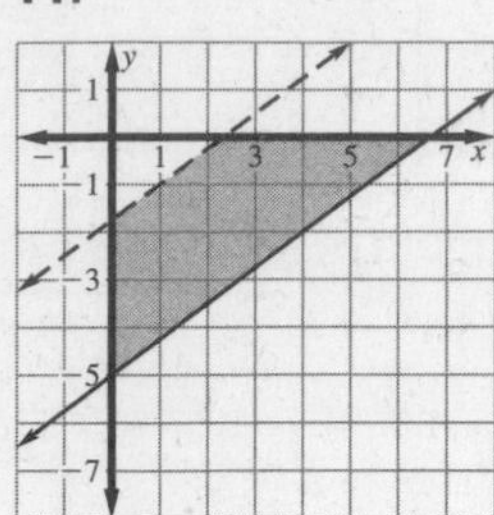

12.

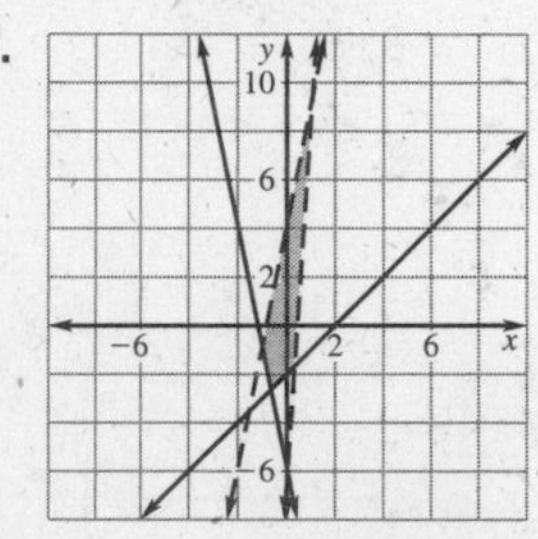

13. (0, 3), (5, −2), (2, 7)

14. (5, 3), (0, 4), (0, −1), (1, 4)

15. (1, 4), (−6, 4), (3, −2), (3, 0)

16.

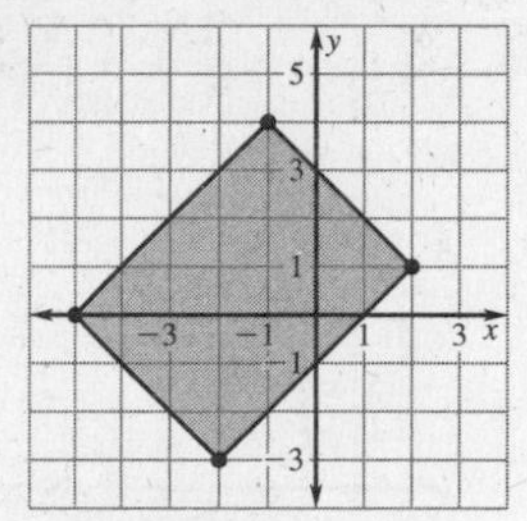

$y \le -x + 3$
$y \ge x - 1$
$y \ge -x - 5$
$y \le x + 5$

17.

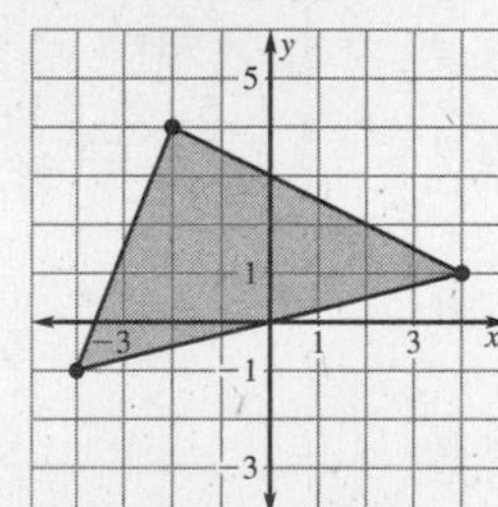

$y \le -\frac{1}{2}x + 3$
$y \le \frac{5}{2}x + 9$
$y \ge \frac{1}{4}x$

18.

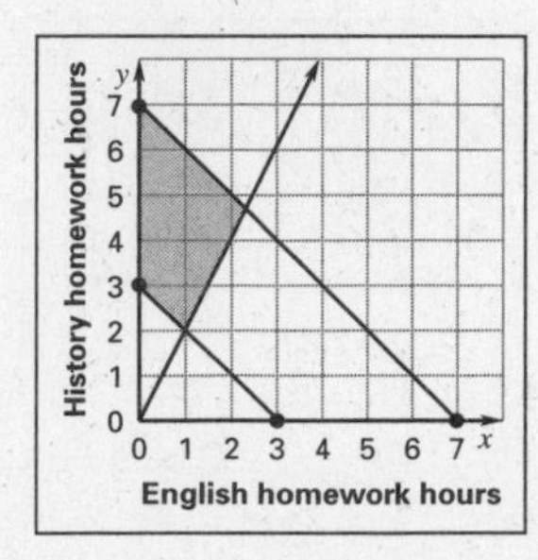

$x + y \ge 3$
$x + y \le 7$
$y \ge 2x$
$x \ge 0$
$y \ge 0$

19.

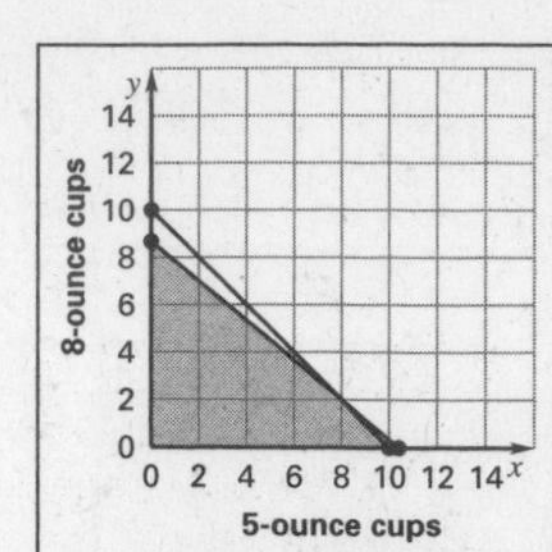

$125x + 150y \le 1300$
$x + y \le 10$
$x \ge 0$
$y \ge 0$

Reteaching with Practice

1.

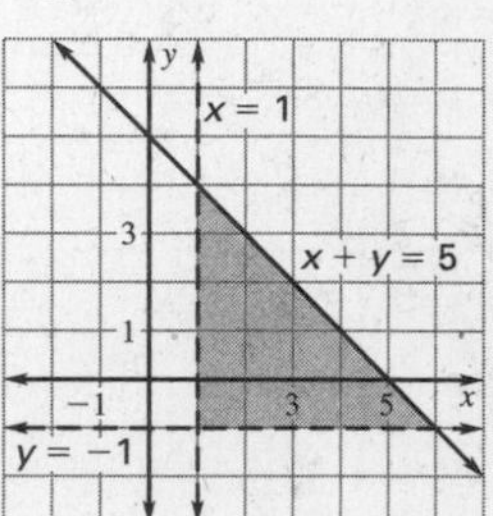

2.

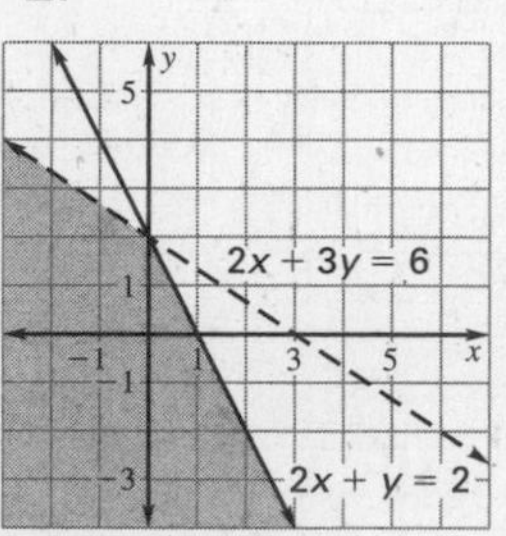

3.

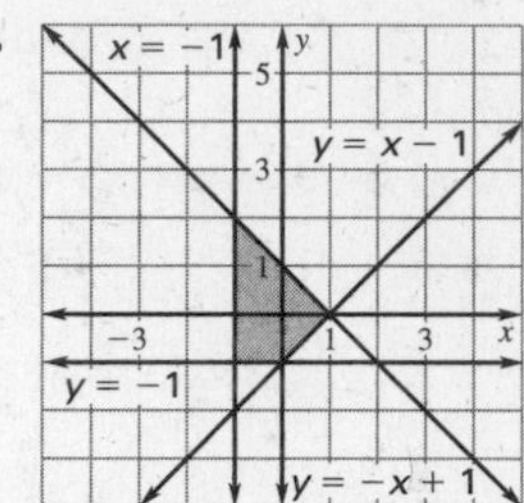

4. $x \ge 0,\ y \ge 0,\ 18x + 9y \le 90$

5. $x \ge 0,\ y \ge 0,\ 16x + 8y \le 72$

Interdisciplinary Application

1. x = number of eels and y = number of fish: $x \le 4, y \le 20, x + y \le 15$

2.

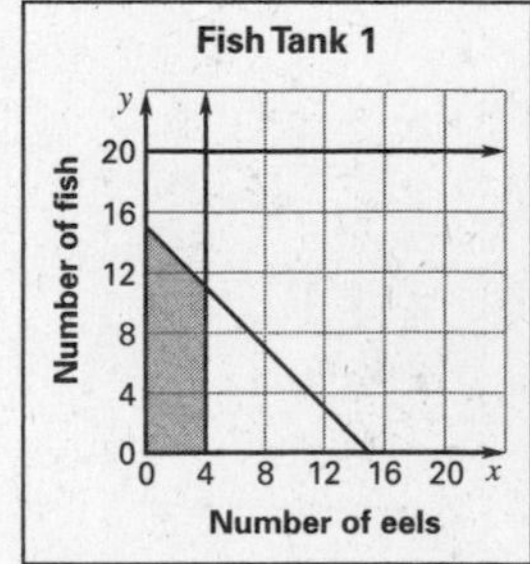

3. There can be 1, 2, or 3 eels in the tank.

4. x = number of lion fish and y = number of parrot fish: $x \geq 5, y \geq 15, x + y \leq 30$

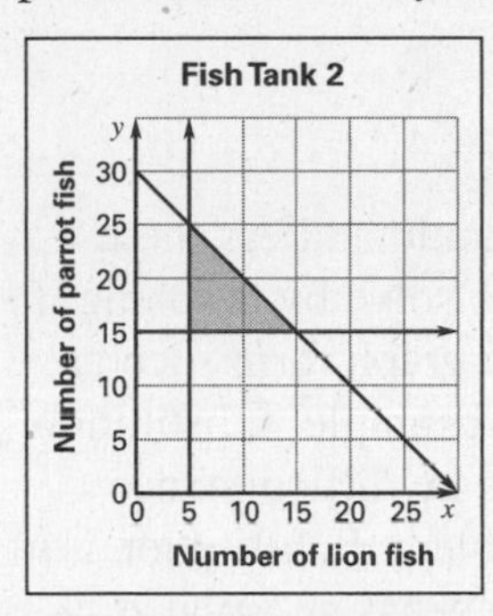

Challenge: Skills and Applications

1.

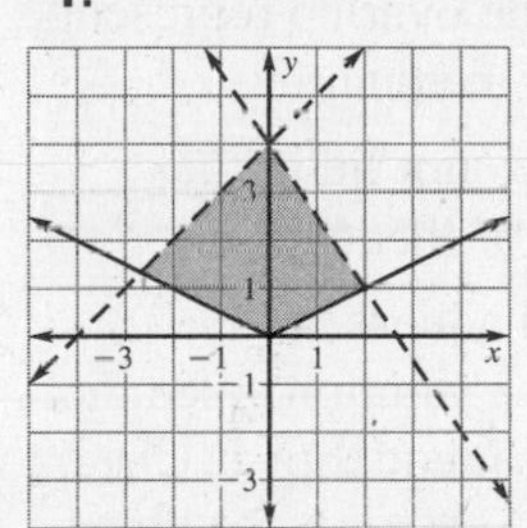

2.

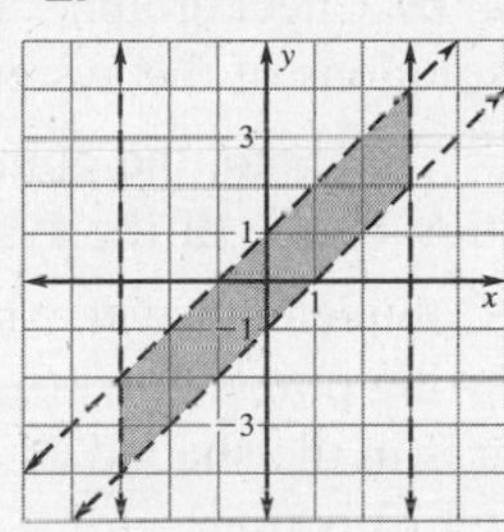

3. $x \geq 0, y \geq 0, 12x + 8y \leq 48, 3x + y \leq 9$

4.

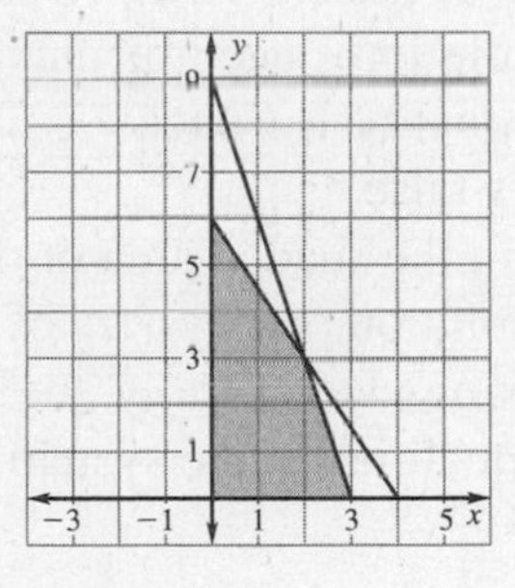

5. 2 large and 3 small dog houses

Quiz 2

1. (2, 2) **2.** infinitely many solutions

3. no solution

4.

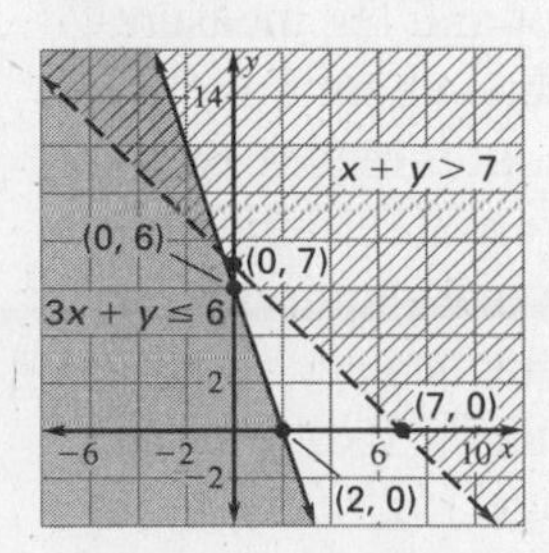

5. $x \geq -6$; $y \leq 3$; $y \geq \frac{1}{2}x + 3$

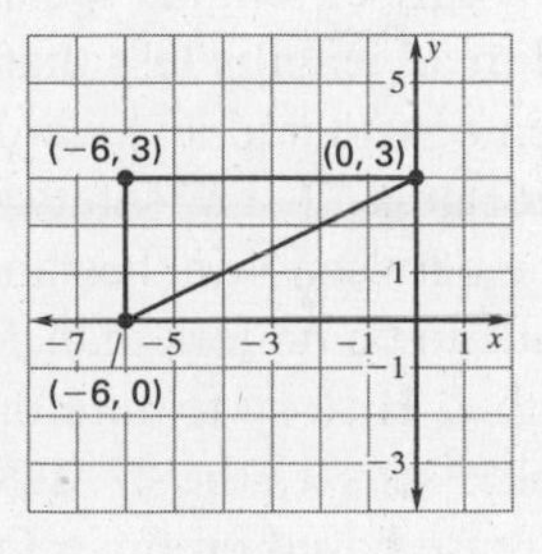

Review and Assessment

Review Games and Activities

1. (3, −7) **2.** All real nos. **3.** (−3, 6) **4.** (7, −1) **5.** (9, −4) **6.** (4, 2) **7.** (5, −8) **8.** No solution **9.** (−6, −5) **10.** No solution **11.** (1, −2) **12.** All real nos.

CRY OVER SPILT MILK

Test A

1. Yes **2.** No

3.

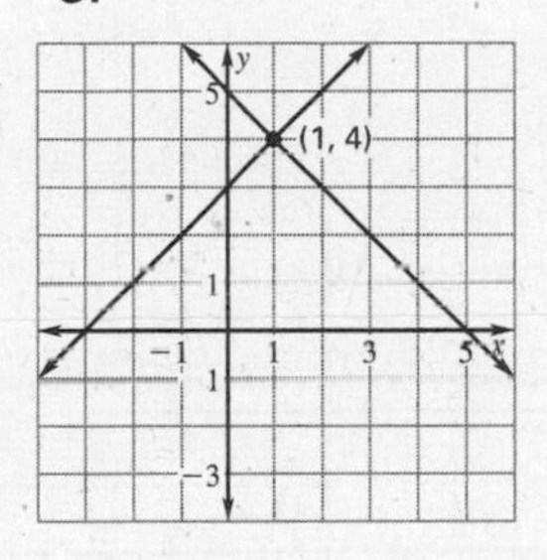

4.

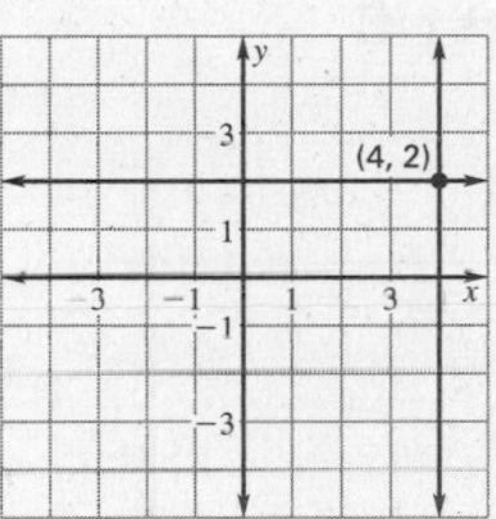

5. (1, 3) **6.** (−1, 4) **7.** \$3: 150 tickets; \$5: 200 tickets **8.** (4, 1) **9.** (1, 4) **10.** 5 **11.** (3, 1); one solution **12.** no solutions **13.** D **14.** A **15.** C **16.** B

17.

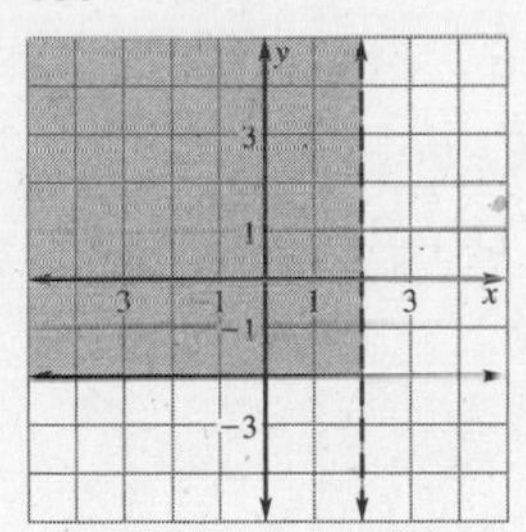

18.

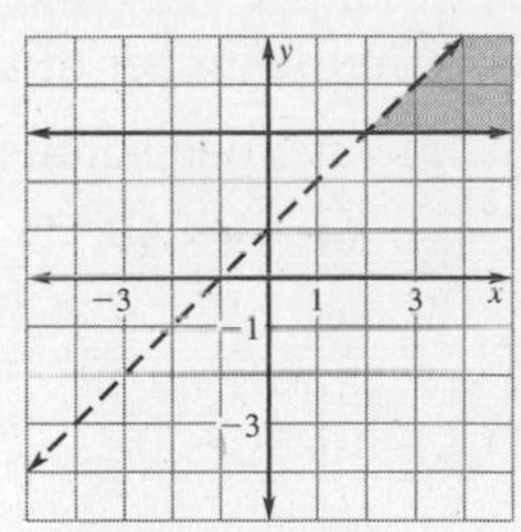

Test B

1. Yes **2.** No

3.

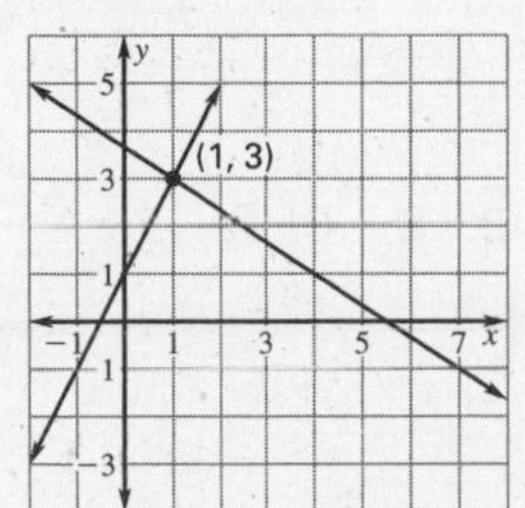

4.

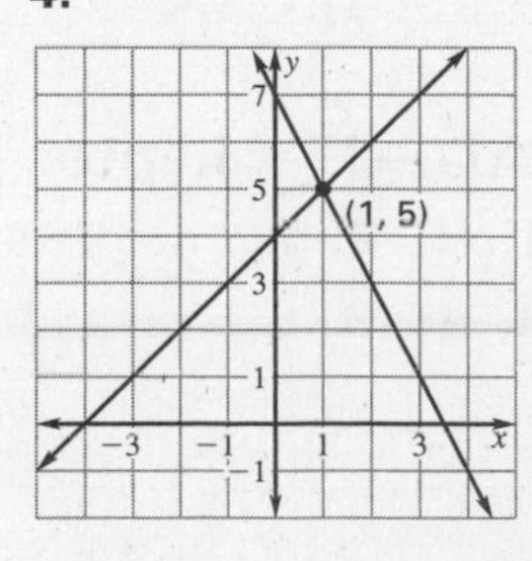

5. (2, 2) **6.** (3, 6) **7.** \$4: 180 tickets; \$6: 270 tickets **8.** (0, 3) **9.** (1, 6) **10.** 4

Review and Assessment *continued*

11. (2, 1); one solution **12.** infinite number of solutions **13.** D **14.** C **15.** B **16.** A

17.

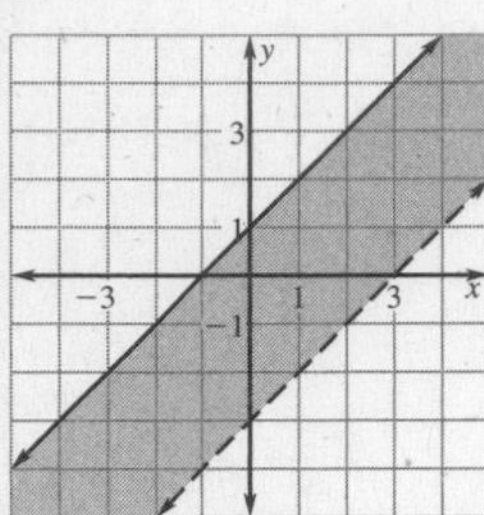

18.

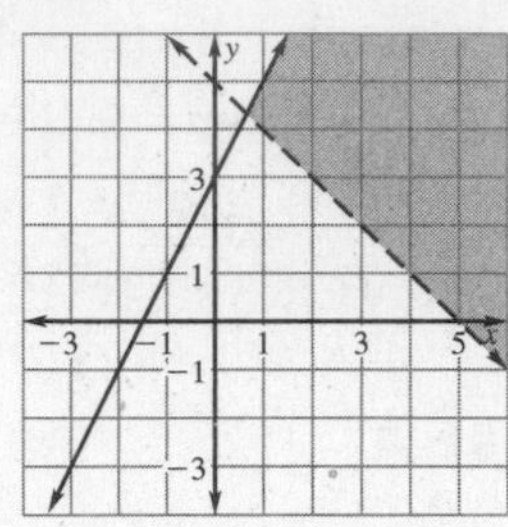

Test C

1.

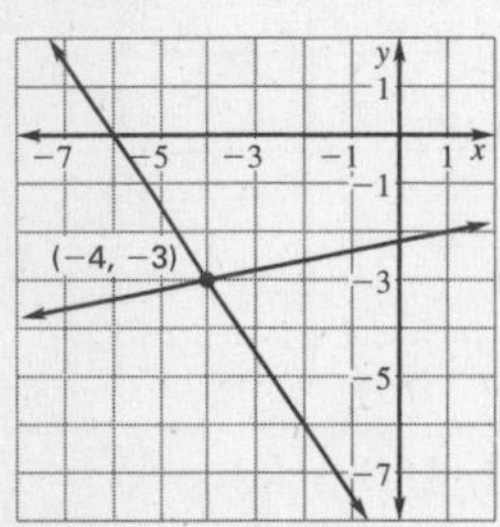

2.

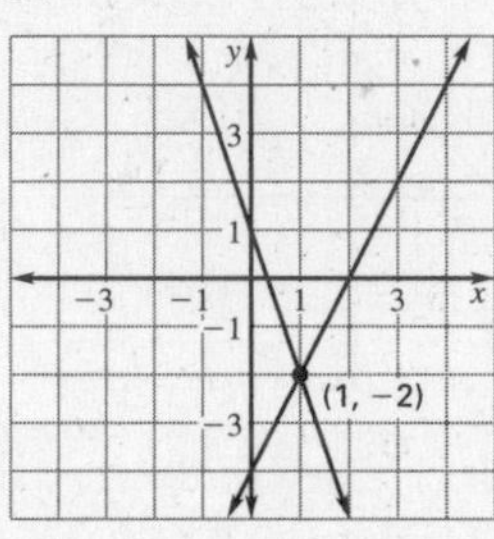

3. $(-5, -10)$ **4.** $(4, -3)$ **5.** A: \$7000; B: \$2000 **6.** $(-3, 2)$ **7.** $(2, 4)$

8. Company A **9.** $(-3, 4)$; one solution

10. infinite number of solutions

11. $(1, -4)$; one solution **12.** no solutions

13. C **14.** A **15.** B **16.** D

17.

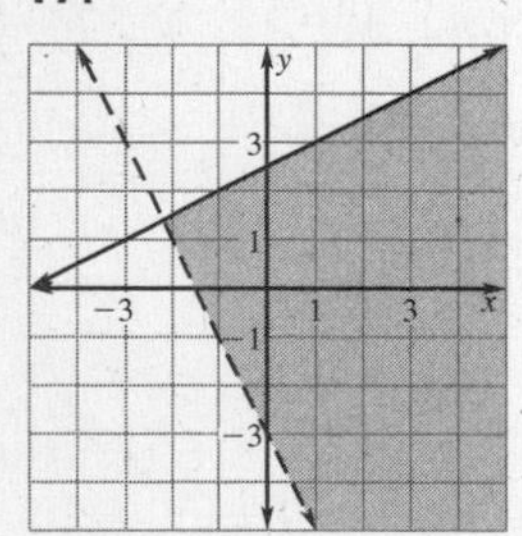

18.

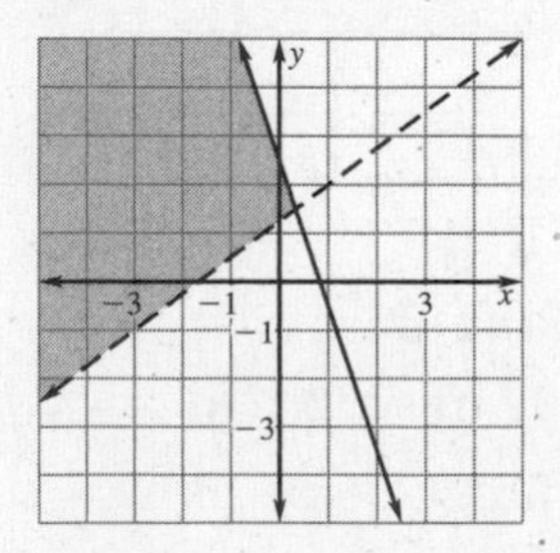

SAT/ACT Chapter Test

1. B **2.** A **3.** B **4.** C **5.** B **6.** D **7.** A **8.** A **9.** C

Alternative Assessment

1. *Sample answers:* **a.** $\begin{cases} y = -2x + 6 \\ y = \frac{1}{3}x - 1 \end{cases}$

b. $\begin{cases} x = 3y - 1 \\ 2x + 4y = 18 \end{cases}$ **c.** $\begin{cases} 3x - 8 = 28 \\ 2x + 4y = 18 \end{cases}$

d. Complete explanations should address these points: A system is easiest to solve by graphing if both equations are in a quick-graph form such as slope-intercept. In the above example, substitution or linear combinations would be difficult since there are fractions in the problem. Substitution is best when one of the equations has an isolated variable. Linear combinations are best if both equations are in standard form.

2. a, b. Check graph **c.** The overlap represents the solutions of the system of inequalities.

d. $\left(\frac{16}{7}, -\frac{13}{7}\right)$, no, the point does not lie in the solution region of the system of inequalities.

3. Complete answers should addess the following points: • (6, 3) is a solution since the point is in the shaded area for Inequality 1. Check: $3 > -13$ is true. • $(-2, -4)$ is not a solution since the point does not lie in the solution region of the graph of Inequality 2. Check: $12 > 12$ is false. • (3, 0) is not a solution since it is only in the shaded region for Inequality 1. Check: $0 > -4$ is true, $6 > 12$ is false.
For a graph, if a point is in the shaded area or on a solid line, the point is a solution of a system of inequalities. When checking algebraically, the point must satisfy all inequalities of the system.

Project: Going Up!

1. Check that student data seems reasonable.

2. Answers will vary. An ideal riser/tread ratio is somewhere between $\frac{1}{2}$ and 1. The range of riser and tread lengths students prefer may vary depending upon their own foot and leg measurements. Riser lengths greater than about 8 inches and tread lengths less than about 9 inches are often considered awkward or unsafe. Some people advocate a "7–11" safety rule that requires risers to be between 4 inches and 7 inches and requires treads to be no less than 11 inches. (Note that the Math & History feature on page 152 of the pupil book discusses the "7–11" rule and contrasts a staircase based on this rule with one based on a formula developed by architect Francois Blondel in 1675.) **3.** Check that tread length and riser length are consistent with answers to Question 2 and that the stairway goes up a total height of 120 inches. Check scale drawings.

4. System of inequalities is;
$r + t \geq 17$
$r + t \leq 18$
$2r + t \geq 24$
$2r + t \leq 25$

Check that students have chosen an appropriate way to label the axes and have graphed the correct boundary lines of the solution region. If the horizontal axis shows tread length and the vertical axis shows riser length, the four vertices of the region are: (9, 8), (10, 7), (11, 7), and (12, 6).

Sample graph:

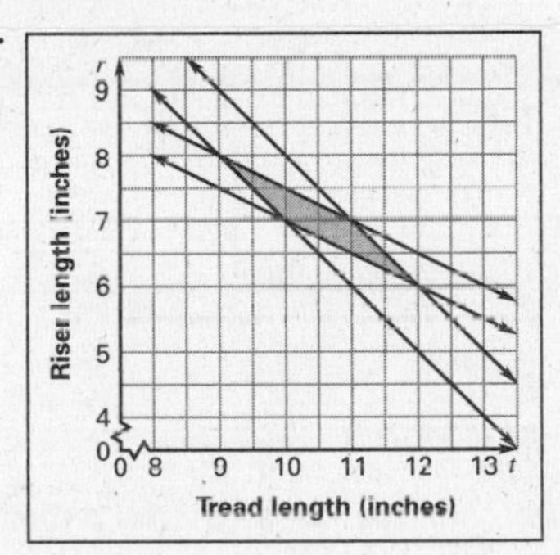

5. Check that points plotted match the data from Questions 1 and 3.

Cumulative Review

1. 738 **2.** 14 **3.** $-\frac{1}{6}$ **4.** -12 **5.** $16b - 10$
6. $-12 + 2s$ **7.** $-6 + \frac{1}{5}x$ **8.** $9.6b - 96$
9. $-4x + 2x^2$ **10.** $2t$ **11.** 6 **12.** 9.89
13. 13 **14.** 0 **15.** 17 **16.** 18 **17.** no
18. yes **19.** yes **20.** yes
21. x-intercept $= -\frac{7}{6}$, y-intercept $= \frac{7}{2}$
22. x-intercept $= 4$, y-intercept $= 1$
23. x-intercept $= 2$, y-intercept $= -\frac{8}{5}$
24. x-intercept $= 6$, y-intercept $= 12$
25. x-intercept $= 35$, y-intercept $= -\frac{7}{12}$
26. x-intercept $= \frac{3}{4}$, y-intercept $= \frac{2}{19}$
27. $9x - y = 10$ **28.** $x + y = 5$
29. $3x - 8y = 0$ **30.** $9x + 10y = -74$
31. $25x - 25y = 86$ **32.** $8x + 4y = 5$

33. $\frac{3}{2} < x < \frac{15}{2}$ **34.** $-10 \leq y \leq 34$
35. $x < 1$ or $x > \frac{3}{2}$ **36.** $x \geq \frac{4}{3}$ or $x \leq -3$
37. $-16.5, 4.5$ **38.** $-22, 23$ **39.** $-1, 5$
40. $-\frac{127}{8}, \frac{129}{8}$

41. $(-1, 7)$

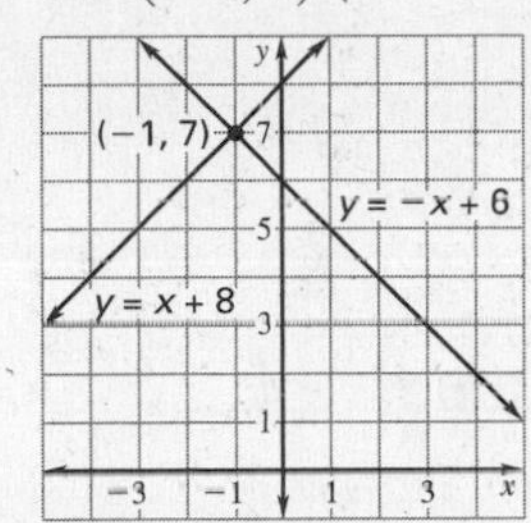

42. $(-1, -4)$

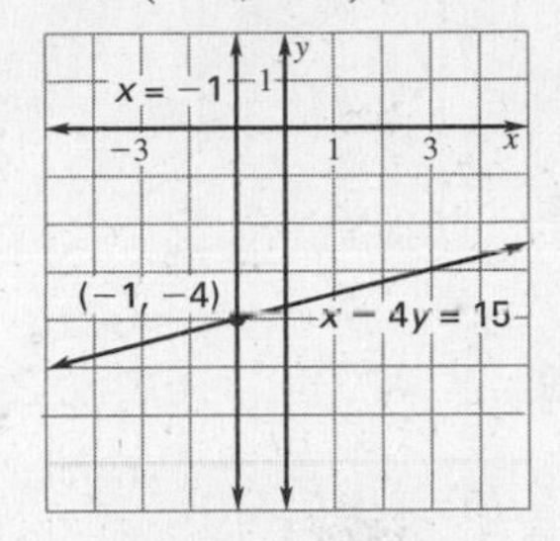

43. $\left(\frac{1}{2}, \frac{7}{2}\right)$

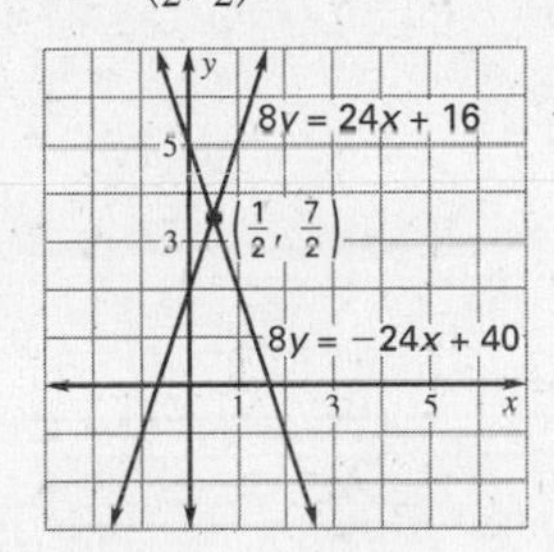

44. $\left(-\frac{10}{3}, \frac{8}{3}\right)$

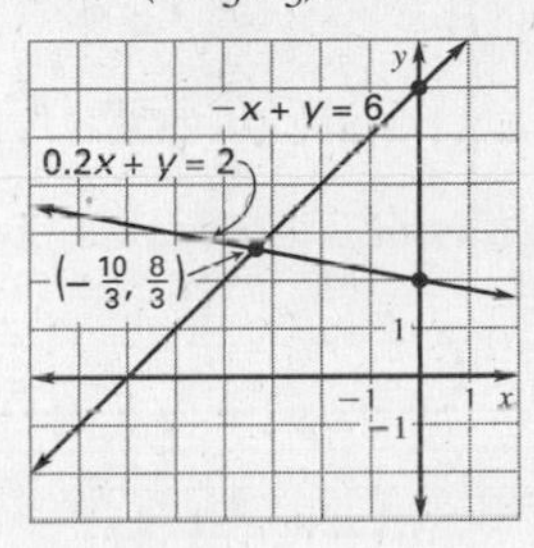

45. (6, 6) **46.** $(-1, -1)$ **47.** $\left(\frac{15}{16}, \frac{3}{16}\right)$
48. about $(1.0054, -0.054)$ **49.** $(-26, 38)$
50. $\left(-\frac{31}{42}, \frac{8}{7}\right)$ **51.** $\left(\frac{1}{3}, -\frac{1}{9}\right)$ **52.** $\left(13, \frac{5}{8}\right)$
53. no solution **54.** infinitely many solutions
55. one solution (0, 10)
56. infinitely many solutions
57. $y < -x + 2, y > 5x - 1$

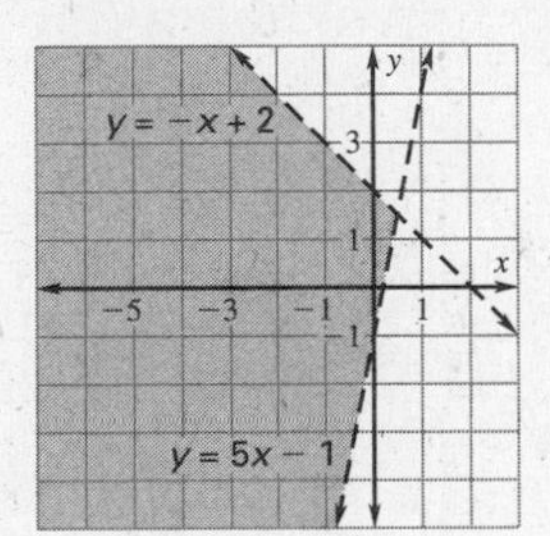

Review and Assessment *continued*

58. $y \geq -x + 8, x < 25$

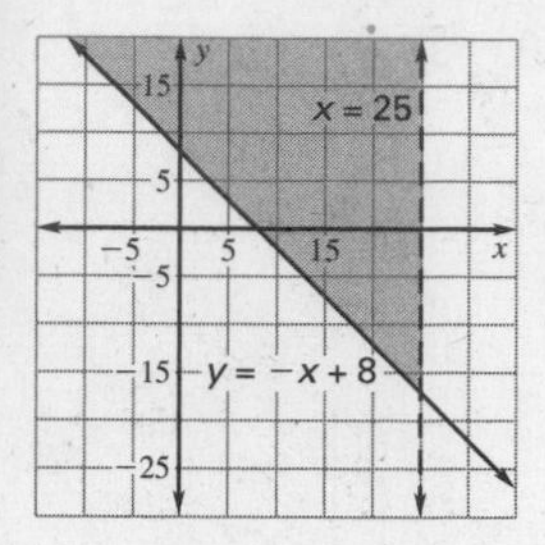

59. $y > -8x + 18, y \geq 0, x \geq 0$

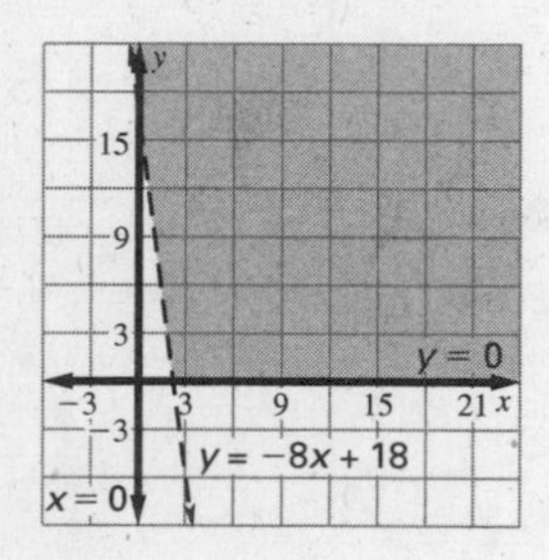

60. $\frac{1}{12}x - \frac{10}{3} \geq y, x - 5 > y$

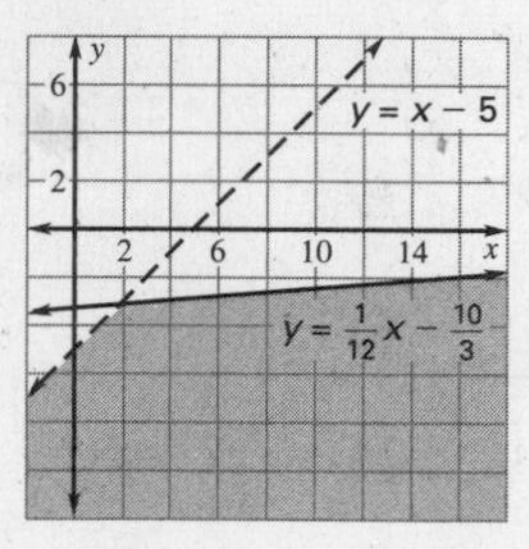